Fabre, Poet of Science by Legros

The Project Gutenberg Etext of Fabre, Poet of Science by Legros Dr. G.V. (C.V.) Legros

Copyright laws are changing all over the world, be sure to check the laws for your country before redistributing these files!!!

Please take a look at the important information in this header.

We encourage you to keep this file on your own disk, keeping an electronic path open for the next readers.

Please do not remove this.

This should be the first thing seen when anyone opens the book. Do not change or edit it without written permission. The words are carefully chosen to provide users with the information they need about what they can legally do with the texts.

Welcome To The World of Free Plain Vanilla Electronic Texts

Etexts Readable By Both Humans and By Computers, Since 1971

These Etexts Prepared By Hundreds of Volunteers and Donations

Information on contacting Project Gutenberg to get Etexts, and further information is included below. We need your donations. The Project Gutenberg Literary Archive Foundation is a 501(c)(3) organization with EIN [Employee Identification Number] 64-6221541

As of 12/12/00 contributions are only being solicited from people in: Colorado, Connecticut, Idaho, Indiana, Iowa, Kentucky, Louisiana, Massachusetts, Montana, Nevada, Oklahoma, South Carolina, South Dakota, Texas, Vermont, and Wyoming.

As the requirements for other states are met, additions to this list will be made and fund raising will begin in the additional states. Please feel free to ask to check the status of your state.

International donations are accepted, but we don't know ANYTHING about how to make them tax-deductible, or even if they CAN be made deductible, and don't have the staff to handle it even if there are ways.

These donations should be made to:

Project Gutenberg Literary Archive Foundation PMB 113 1739 University Ave. Oxford, MS 38655-4109

Title: Fabre, Poet of Science

Author: Dr. G.V. (C.V.) Legros

Release Date: October, 2002 [Etext #3489] [Yes, we are about one year ahead of schedule] [The actual date this file first posted = 05/20/01]

Edition: 10

Language: English

The Project Gutenberg Etext of Fabre, Poet of Science by Legros ******This file should be named fbrps10.txt or fbrps10.zip*****

Corrected EDITIONS of our etexts get a new NUMBER, fbrps11.txt VERSIONS based on separate sources get new LETTER, fbrps10a.txt

This etext was produced by Sue Asscher <asschers@dingoblue.net.au>

Project Gutenberg Etexts are usually created from multiple editions, all of which are in the Public Domain in the United States, unless a copyright notice is included. Therefore, we usually do NOT keep any of these books in compliance with any particular paper edition.

We are now trying to release all our books one year in advance of the official release dates, leaving time for better editing. Please be encouraged to send us error messages even years after the official publication date.

Please note: neither this list nor its contents are final till midnight of the last day of the month of any such announcement. The official release date of all Project Gutenberg Etexts is at Midnight, Central Time, of the last day of the stated month. A preliminary version may often be posted for suggestion, comment and editing by those who wish to do so.

Most people start at our sites at: http://gutenberg.net
http://promo.net/pg

Those of you who want to download any Etext before announcement can surf to them as follows, and just download by date; this is also a good way to get them instantly upon announcement, as the indexes our cataloguers produce obviously take a while after an announcement goes out in the Project Gutenberg Newsletter.

http://www.ibiblio.org/gutenberg/etext02 or
ftp://ftp.ibiblio.org/pub/docs/books/gutenberg/etext02

Or /etext01, 00, 99, 98, 97, 96, 95, 94, 93, 92, 92, 91 or 90

Just search by the first five letters of the filename you want, as it appears in our Newsletters.

Information about Project Gutenberg

(one page)

We produce about two million dollars for each hour we work. The time it takes us, a rather conservative estimate, is fifty hours to get any etext selected, entered, proofread, edited, copyright searched and analyzed, the copyright letters written, etc. This projected audience is one hundred million readers. If our value per text is nominally estimated at one dollar then we produce $2 million dollars per hour this year as we release fifty new Etext files per month, or 500 more Etexts in 2000 for a total of 3000+ If they reach just 1-2% of the world's population then the total should reach over 300 billion Etexts given away by year's end.

The Goal of Project Gutenberg is to Give Away One Trillion Etext Files by December 31, 2001. [10,000 x 100,000,000 = 1 Trillion] This is ten thousand titles each to one hundred million readers, which is only about 4% of the present number of computer users.

At our revised rates of production, we will reach only one-third of that goal by the end of 2001, or about 3,333 Etexts unless we manage to get some real funding.

The Project Gutenberg Literary Archive Foundation has been created to secure a future for Project Gutenberg into the next millennium.

We need your donations more than ever!

Presently, contributions are only being solicited from people in: Colorado, Connecticut, Idaho, Indiana, Iowa, Kentucky, Louisiana, Massachusetts, Montana, Nevada, Oklahoma, South Carolina, South Dakota, Texas, Vermont, and Wyoming.

As the requirements for other states are met, additions to this list will be made and fund raising will begin in the additional states.

These donations should be made to:

Project Gutenberg Literary Archive Foundation PMB 113 1739 University Ave. Oxford, MS 38655-4109

Project Gutenberg Literary Archive Foundation, EIN [Employee Identification Number] 64-6221541, has been approved as a 501(c)(3) organization by the US Internal Revenue Service (IRS). Donations are tax-deductible to the extent permitted by law. As the requirements for other states are met, additions to this list will be made and fund raising will begin in the additional states.

All donations should be made to the Project Gutenberg Literary Archive Foundation. Mail to:

Project Gutenberg Literary Archive Foundation PMB 113
1739 University Avenue Oxford, MS 38655-4109 [USA]

We need your donations more than ever!

You can get up to date donation information at:

http://www.gutenberg.net/donation.html

*** If you can't reach Project Gutenberg, you can always email directly to:

Michael S. Hart <hart@pobox.com>

hart@pobox.com forwards to hart@prairienet.org and archive.org if your mail bounces from archive.org, I will still see it, if it bounces from prairienet.org, better resend later on. . . .

Prof. Hart will answer or forward your message.

We would prefer to send you information by email.

***

Example command-line FTP session:

```
ftp ftp.ibiblio.org
login: anonymous
password: your@login
cd pub/docs/books/gutenberg
cd etext90 through etext99 or etext00 through etext02, et
dir [to see files]
get or mget [to get files. . .set bin for zip files]
GET GUTINDEX.??  [to get a year's listing of books, e.g.,
GUTINDEX.99]
GET GUTINDEX.ALL [to get a listing of ALL books]
```

**

The Legal Small Print

**

(Three Pages)

START**THE SMALL PRINT!**FOR PUBLIC DOMAIN ETEXTS**START Why is this "Small Print!" statement here? You know: lawyers. They tell us you might sue us if there is something wrong with your copy of this etext, even if you got it for free from someone other than us, and even if what's wrong is not our fault. So, among other things, this "Small Print!" statement disclaims most of our liability to you. It also tells you how you may distribute copies of this etext if you want to.

BEFORE! YOU USE OR READ THIS ETEXT

By using or reading any part of this PROJECT GUTENBERG-tm etext, you indicate that you understand, agree to and accept this "Small Print!" statement. If you do not, you can receive a refund of the money (if any) you paid for this etext by sending a request within 30 days of receiving it to the person you got it from. If you received this etext on a physical medium (such as a disk), you must return it with your request.

ABOUT PROJECT GUTENBERG-TM ETEXTS

This PROJECT GUTENBERG-tm etext, like most PROJECT GUTENBERG-tm etexts, is a "public domain" work distributed by Professor Michael S. Hart through the Project Gutenberg Association (the "Project"). Among other things, this means that no one owns a United States copyright on or for this work, so the Project (and you!) can copy and distribute it in the United States without permission and without paying copyright royalties. Special rules, set forth below, apply if you wish to copy and distribute this etext under the "PROJECT GUTENBERG" trademark.

Please do not use the "PROJECT GUTENBERG" trademark to market any commercial products without permission.

To create these etexts, the Project expends considerable efforts to identify, transcribe and proofread public domain works. Despite these efforts, the Project's etexts and any medium they may be on may contain "Defects". Among other things, Defects may take the form of incomplete, inaccurate or corrupt data, transcription errors, a copyright or other intellectual property infringement, a defective or damaged disk or other etext medium, a computer virus, or computer codes that damage or cannot be read by your equipment.

LIMITED WARRANTY; DISCLAIMER OF DAMAGES

But for the "Right of Replacement or Refund" described below, [1] Michael Hart and the Foundation (and any other party you may receive this etext from as a PROJECT GUTENBERG-tm etext) disclaims all liability to you for damages, costs and expenses, including legal fees, and [2] YOU HAVE NO REMEDIES FOR NEGLIGENCE OR

UNDER STRICT LIABILITY, OR FOR BREACH OF WARRANTY OR CONTRACT, INCLUDING BUT NOT LIMITED TO INDIRECT, CONSEQUENTIAL, PUNITIVE OR INCIDENTAL DAMAGES, EVEN IF YOU GIVE NOTICE OF THE POSSIBILITY OF SUCH DAMAGES.

If you discover a Defect in this etext within 90 days of receiving it, you can receive a refund of the money (if any) you paid for it by sending an explanatory note within that time to the person you received it from. If you received it on a physical medium, you must return it with your note, and such person may choose to alternatively give you a replacement copy. If you received it electronically, such person may choose to alternatively give you a second opportunity to receive it electronically.

THIS ETEXT IS OTHERWISE PROVIDED TO YOU "AS-IS". NO OTHER WARRANTIES OF ANY KIND, EXPRESS OR IMPLIED, ARE MADE TO YOU AS TO THE ETEXT OR ANY MEDIUM IT MAY BE ON, INCLUDING BUT NOT LIMITED TO WARRANTIES OF MERCHANTABILITY OR FITNESS FOR A PARTICULAR PURPOSE.

Some states do not allow disclaimers of implied warranties or the exclusion or limitation of consequential damages, so the above disclaimers and exclusions may not apply to you, and you may have other legal rights.

INDEMNITY

You will indemnify and hold Michael Hart, the Foundation, and its trustees and agents, and any volunteers associated with the production and distribution of Project Gutenberg-tm texts harmless, from all liability, cost and expense, including legal fees, that arise directly or indirectly from any of the following that you do or cause: [1] distribution of this etext, [2] alteration, modification, or addition to the etext, or [3] any Defect.

DISTRIBUTION UNDER "PROJECT GUTENBERG-tm"

You may distribute copies of this etext electronically, or by disk, book or any other medium if you either delete this "Small Print!" and all other references to Project Gutenberg, or:

[1] Only give exact copies of it. Among other things, this requires that you do not remove, alter or modify the etext or this "small print!" statement. You may however, if you wish, distribute this etext in machine readable binary, compressed, mark-up, or proprietary form, including any form resulting from conversion by word processing or hypertext software, but only so long as *EITHER*:

[*] The etext, when displayed, is clearly readable, and does *not* contain characters other than those intended by the author of the work, although tilde (~), asterisk (*) and underline (_i_) characters may be used to convey punctuation intended by the author, and additional characters may be used to indicate hypertext links; OR

[*] The etext may be readily converted by the reader at no expense into plain ASCII, EBCDIC or equivalent form by the program that displays the etext (as is the case, for instance, with most word processors); OR

[*] You provide, or agree to also provide on request at no additional cost, fee or expense, a copy of the etext in its original plain ASCII form (or in EBCDIC or other equivalent proprietary form).

[2] Honor the etext refund and replacement provisions of this "Small Print!" statement.

[3] Pay a trademark license fee to the Foundation of 20% of the gross profits you derive calculated using the method you already use to calculate your applicable taxes. If you don't derive profits, no royalty is due. Royalties are payable to "Project Gutenberg Literary Archive Foundation" the 60 days following each date you prepare (or were legally required to prepare) your annual (or equivalent periodic)

tax return. Please contact us beforehand to let us know your plans and to work out the details.

WHAT IF YOU *WANT* TO SEND MONEY EVEN IF YOU DON'T HAVE TO?

Project Gutenberg is dedicated to increasing the number of public domain and licensed works that can be freely distributed in machine readable form.

The Project gratefully accepts contributions of money, time, public domain materials, or royalty free copyright licenses. Money should be paid to the: "Project Gutenberg Literary Archive Foundation."

If you are interested in contributing scanning equipment or software or other items, please contact Michael Hart at: hart@pobox.com

*END THE SMALL PRINT! FOR PUBLIC DOMAIN ETEXTS*Ver.12.12.00*END*

This etext was produced by Sue Asscher <asschers@dingoblue.net.au>

FABRE, POET OF SCIENCE

by DR. G.-V. LEGROS.

"De fimo ad excelsa." J.-H. Fabre.

WITH A PREFACE BY JEAN-HENRI FABRE.

TRANSLATED BY BERNARD MIALL.

PREFACE.

The good friend who has so successfully terminated the task which he felt a vocation to undertake thought it would be of advantage to complete it by presenting to the reader a picture both of my life as a whole and of the work which it has been given me to accomplish.

The better to accomplish his undertaking, he abstracted from my correspondence, as well as from the long conversations which we have so often enjoyed together, a great number of those memories of varying importance which serve as landmarks in life; above all in a life like mine, not exempt from many cares, yet not very fruitful in incidents or great vicissitudes, since it has been passed very largely, in especial during the last thirty years, in the most absolute retirement and the completest silence.

Moreover, it was not unimportant to warn the public against the errors, exaggerations, and legends which have collected about my person, and thus to set all things in their true light.

In undertaking this task my devoted disciple has to some extent been able to replace those "Memoirs" which he suggested that I should write, and which only my bad health has prevented me from undertaking; for I feel that henceforth I am done with wide horizons and "far-reaching thoughts."

And yet on reading now the old letters which he has exhumed from a mass of old yellow papers, and which he has presented and co-ordinated with so pious a care, it seems to me that in the depths of my being I can still feel rising in me all the fever of my early years, all the enthusiasm of long ago, and that I should still be no less ardent a worker were not the weakness of my eyes and the failure of my strength to-day an insurmountable obstacle.

Thoroughly grasping the fact that one cannot write a biography without entering into the sphere of those ideas which alone make a life interesting, he has revived around me that world which I have so long contemplated, and summarized in a striking epitome, and as a strict interpreter, my methods (which are, as will be seen, within the reach of all), my ideas, and the whole body of my works and discoveries; and despite the obvious difficulty which such an attempt would appear to present, he has succeeded most wonderfully in achieving the most lucid, complete, and vital exposition of these matters that I could possibly have wished.

Jean-Henri Fabre.

Sérignan, Vaucluse, November 12, 1911.

CONTENTS.

PREFACE.

INTRODUCTION.

CHAPTER 1.

THE INTUITION OF NATURE.

CHAPTER 2.

THE PRIMARY TEACHER.

CHAPTER 3.

CORSICA.

CHAPTER 4.

AT AVIGNON.

CHAPTER 5.

A GREAT TEACHER.

CHAPTER 6.

THE HERMITAGE.

CHAPTER 7.

THE INTERPRETATION OF NATURE.

CHAPTER 8.

THE MIRACLE OF INSTINCT.

CHAPTER 9.

EVOLUTION OR "TRANSFORMISM."

CHAPTER 10.

THE ANIMAL MIND.

CHAPTER 11.

HARMONIES AND DISCORDS.

CHAPTER 12.

THE TRANSLATION OF NATURE.

CHAPTER 13.

THE EPIC OF ANIMAL LIFE.

CHAPTER 14.

PARALLEL LIVES.

CHAPTER 15.

THE EVENINGS AT SÉRIGNAN.

CHAPTER 16.

TWILIGHT.

NOTES.

INDEX.

INTRODUCTION.

Here I offer to the public the life of Jean-Henri Fabre; at once an admiring commentary upon his work and an act of pious homage, such as ought to be offered, while he lives, to the great naturalist who is even to-day so little known.

Hitherto it was not easy to speak of Henri Fabre with exactitude. An enemy to all advertisement, he has so discreetly held himself withdrawn that one might almost say that he has encouraged, by his silence, many doubtful or unfounded rumours, which in course of time would become even more incorrect.

For example, although quite recently his material situation was presented in the gloomiest of lights, while it had really for some time ceased to be precarious, it is none the less true that during his whole life he has had to labour prodigiously in order to earn a little money to feed and rear his family, to the great detriment of his scientific inquiries; and we cannot but regret that he was not freed from all material cares at least twenty years earlier than was the case.

But he was not one to speak of his troubles to the first comer; and it was only after the sixth volume of the "Souvenirs entomologiques" had appeared that his reserve was somewhat mitigated. Yet it was necessary that he should speak of these troubles, that he should tell everything; and, thanks to his conversation and his letters, I have been able to revive the past.

Among the greatest of my pleasures I count the notable honour of having known him, and intimately. As an absorbed and attentive witness I was present at the accomplishment of his last labours; I watched his last years of work, so critical, so touching, so forsaken, before his ultimate resurrection. What fruitful and suggestive lessons I learned in his company, as we paced the winding paths of his Harmas; or while I sat beside him, at his patriarchal table, interrogating that memory of his, so rich in remembrances that even the remotest events of his life were as near to him as those that had only then befallen him; so that the majority of the judgments to be found in this book, of which not a line has been written without his approval, may be regarded as the direct emanation of his mind.

As far as possible I have allowed him to speak himself. Has he not sketched the finest pages of his "biography of a solitary student" in those racy chapters of his "Souvenirs": those in which he has developed his genesis as a naturalist and the history of the evolution of his ideas? (Introduction/1.) In all cases I have only introduced such indications as were essential to complete the sequence of events. It would have been idle to re-tell in the same terms what every one may read elsewhere, or to repeat in different and less happy terms what Fabre himself has told so well.

I have therefore applied myself more especially to filling the gaps which he has left, by listening to his conversation, by appealing to his memories, by questioning his contemporaries, by recording the impressions of his sometime pupils. I have endeavoured to assemble all these data, in order to authenticate them, and have also gleaned many facts among his manuscripts (Introduction/2.), and have had recourse to all that portion of his correspondence which fortunately fell into my hands.

This correspondence, to be truthful, does not appear at any time to have been very assiduous. Fabre, as we shall see in the story of his life (Introduction/3.), disliked writing letters, both in his studious youth and during the later period of isolation and silence.

On the other hand, although he wrote but little, he never wrote with difficulty or as a mere matter of duty. Among all the letters which I have succeeded in collecting there are scarcely any that are not of interest from one point of view or another. No frivolous narratives, no futile acquaintances, no commonplace intimacies; everything in his life is serious, and everything makes for a goal.

But we must set apart, as surpassing all others in interest, the letters which Fabre addressed to his brother during the years spent as schoolmaster at Carpentras or Ajaccio; for these are more especially instructive in respect of the almost unknown years of his youth; these most of all reveal his personality and are one of the finest illustrations that could be given of his life, a true poem of energy and disinterested labour.

I have to thank M. Frédéric Fabre, who, in his fraternal piety, has generously placed all his family records at my disposal, and also his two sons, my dear friends Antonin Fabre, councillor at the Court of Nîmes, and Henri Fabre, of Avignon, for these precious documents; and I take this opportunity of expressing my profound gratitude.

CHAPTER 16.

Let me at the same time thank all those who have associated themselves with my efforts by supplying me with letters in their possession and furnishing me with personal information; and in particular Mme Henry Devillario, M. Achard, and M. J. Belleudy, ex-prefect of Vaucluse; not forgetting M. Louis Charrasse, teacher at Beaumont-d'Orange, and M. Vayssières, professor of the Faculty of Sciences at Marseilles, all of whom I have to thank for personal and intimate information.

I must also express my gratitude to M. Henri Bergson, Professor Bouvier, and the learned M. Paul Marchal for the advice and the valuable suggestions which they offered me during the preparation of this book.

I shall feel fully repaid for my pains if this "Life" of one of the greatest of the world's naturalists, by enabling men to know him better, also leads them to love him the more.

FABRE, POET OF SCIENCE.

CHAPTER 1.

THE INTUITION OF NATURE.

Each thing created, says Emerson, has its painter or its poet. Like the enchanted princess of the fairy-tales, it awaits its predestined liberator.

Every part of nature has its mystery and its beauty, its logic and its explanation; and the epigraph given me by Fabre himself, which appears on the title-page of this volume, is in no way deceptive. The tiny insects buried in the soil or creeping over leaf or blade have for him been sufficient to evoke the most important, the most fascinating problems, and have revealed a whole world of miracle and poetry.

He saw the light at Saint-Léons, a little commune of the canton of Vezins in the Haut Rouergue, on the 22nd December, 1823, some seven years earlier than Mistral, his most famous neighbour, the greater lustre of whose celebrity was to eclipse his own.

Here he essayed his earliest steps; here he stammered his first syllables.

His early childhood, however, was passed almost wholly at Malaval, a tiny hamlet in the parish of Lavaysse, whose

CHAPTER 1.

belfry was visible at quite a short distance; but to reach it one had to travel nearly twenty-five rough, mountainous miles, through a whole green countryside; green, but bare, and lacking in charm. (1/1.)

All his paternal forebears came from Malaval, and thence one day his father, Antoine Fabre, came to dwell at Saint-Léons, as a consequence of his marriage with the daughter of the huissier, Victoire Salgues, and in order to prepare himself, as working apprentice, in the tricks and quibbles of the law. (1/2.)

In the roads of Malaval, bordered with brambles, in the glades of bracken, and amid the meadows of broom, he received his first impressions of nature. At Malaval too lived his grandmother, the good old woman who could lull him to sleep at night with beautiful stories and simple legends, while she wound her distaff or spun her bobbin.

But what were all these imaginary marvels, what were the ogres who smelt fresh meat, or "the fairies who turned pumpkins into coaches and lizards into footmen" beside all the marvels of reality, which already he was beginning to perceive?

For above all things he was born a poet: a poet by instinct and by vocation. From his earliest childhood, "the brain hardly released from the swaddling-bands of unconsciousness," the things of the outer world left a profound and living impression. As far back as he can remember, while still quite a child, "a little monkey of six, still dressed in a little baize frock," or just "wearing his first braces," he sees himself "in ecstasy before the splendours of the wing-cases of a gardener-beetle, or the wings of a butterfly." At nightfall, among the bushes, he learned to recognize the chirp of the grasshopper. To put it in his own words, "he made for the flowers and insects as the Pieris makes for the cabbage and the Vanessa makes for the nettle." The riches of the rocks; the life which swarms in the depth of the waters; the world of plants and animals, that "prodigious poem; all nature filled him with curiosity and wonder." "A voice charmed him; untranslatable; sweeter than language and vague as a dream." (1/3.)

These peculiarities are all the more astonishing in that they seem to be absolutely spontaneous and in nowise hereditary. What his parents were he himself has told us: small farmers, cultivating a little unprofitable land; poor "husbandmen, sowers of rye, cowherds"; and in the wretched surroundings of his childhood, when the only light, of an evening, came from a splinter of pine, steeped in resin, which was held by a strip of slate stuck into the wall; when his folk shut themselves in the byre, in times of severe cold, to save a little firewood and while away the evenings; when close at hand, through the bitter wind, they heard the howling of the wolves: here, it would seem, was nothing propitious to the birth of such tastes, if he had not borne them naturally within him.

But is it not the very essence of genius, as it is the peculiarity of instinct, to spring from the depths of the invisible?

Yet who shall say what stores of thought unspoken, what unknown treasures of observation never to be communicated, what patient reflections unuttered, may be housed in those toil-worn brains, in which, perhaps, slowly and obscurely, accumulate the germs of faculties and talents by which some more favoured descendant may one day benefit? How many poets have died unpublished or unperceived, in whom only the power of expression was lacking!

When he was seven years old his parents recalled him to Saint-Léons, in order to send him to the school kept by his godfather, Pierre Ricard, the village schoolmaster, "at once barber, bellringer, and singer in the choir." Rembrandt, Teniers, nor Van Ostade never painted anything more picturesque than the room which served at the same time as kitchen, refectory, and bedroom, with "halfpenny prints papering the walls" and "a huge chimney, for which each

CHAPTER 1.

had to bring his log of a morning in order to enjoy the right to a place at the fireside."

He was never to forget these beloved places, blessed scenes of his childhood, amid which he grew up like a little savage, and through all his material sufferings, all his hours of bitterness, and even in the resignation of age, their idyllic memory sufficed to make his life fragrant. He would always see the humble paternal garden, the brook where he used to surprise the crayfish, the ash-tree in which he found his first goldfinch's nest, and "the flat stone on which he heard, for the first time, the mellow ringing of the bellringer frog." (1/4.) Later, when writing to his brother, he was to recall the good days of still careless life, when "he would sprawl, the sun on his belly, on the mosses of the wood of Vezins, eating his black bread and cream" or "ring the bells of Saint-Léons" and "pull the tails of the bulls of Lavaysse." (1/5.)

For Henri had a brother, Frédéric, barely two years younger than he; equally meditative by nature, and of a serious, upright mind; but his tastes inclined rather to matters of administration and the understanding of business, so that where Frédéric was bored, Henri was more than content, thirstily drinking in science and poetry "among the blue campanulas of the hills, the pink heather of the mountains, the golden buttercups of the meadows,

CHAPTER 1.

and the odorous bracken of the woods." (1/6.) Apart from this the two brothers "were one"; they understood one another in a marvellous fashion, and always loved one another. Henri never failed to watch over Frédéric with a wholly fatherly solicitude; he was prodigal of advice, helpful with his experience, doing his best to smooth away all difficulties, encouraging him to walk in his footsteps and make his way through the world behind him. He was his confidant, giving an ear to all that befell him of good or ill; to his fears, his disappointments, his hopes, and all his thoughts; and he took the keenest interest in his studies and researches. On the other hand, he had no more sure and devoted friend; none more proud of his first success, and in later days no more enthusiastic admirer, and none more eager for his fame. (1/7.)

He was twelve years old when his father, "the first of all his line, was tempted by the town," and led all his family to Rodez, there to keep a café. The future naturalist entered the school of this town, where he served Mass on Sunday, in the chapel, in order to pay his fees. There again he was interested in the animal creation above all. When he began to construe Virgil the only thing that charmed him, and which he remembered, was the landscape in which the persons of the poem move, in which are so many "exquisite details concerning the cicada, the goat, and the laburnum."

CHAPTER 1.

Thus four years went by: but then his parents were constrained to seek their fortune elsewhere, and transported their household to Toulouse, where again the father kept a café. The young Henri was admitted gratuitously to the seminary of the Esquille, where he managed to complete his fifth year. Unfortunately his progress was soon interrupted by a new exodus on the part of his family, which emigrated this time to Montpellier, where he was haunted for a time by dreams of medicine, to which he seemed notably adapted. Finally, a run of bad luck persisting, he had to bid farewell to his studies and gain his bread as best he could. We see him set out along the wide white roads: lost, almost a wanderer, seeking his living by the sweat of his brow; one day selling lemons at the fair of Beaucaire, under the arcades of the market or before the barracks of the Pré; another day enlisting in a gang of labourers who were working on the line from Beaucaire to Nîmes, which was then in process of construction. He knew gloomy days, lonely and despairing. What was he doing? of what was he dreaming? The love of nature and the passion for learning sustained him in spite of all, and often served him as nourishment; as on the day when he dined on a few grapes, plucked furtively at the edge of a field, after exchanging the poor remnant of his last halfpence for a little volume of Reboul's poems; soothing his hunger by reciting the verses of the gentle baker-poet. Often some creature kept him company; some insect never seen before was often his greatest pleasure; such as the pine-chafer, which he encountered then for the first time; that superb beetle, whose black or chestnut coat is sprinkled with specks of white velvet; which squeaks when captured, emitting a slight complaining sound, like the vibration of a pane of glass rubbed with the tip of a moistened finger. (1/8.)

Already this young mind, romantic and classic at once, full of the ideal, and so positive that it seemed to seek support in an intense grasp of things and beings--two gifts well-nigh incompatible, and often mutually destructive--already it knew, not only the love of study and a passion for the truth, but the sovereign delight of feeling everything and understanding everything.

It was under these conditions--that is, amid the rudest privations--that he ventured to enter a competitive examination for a bursary at the École Normale Primaire of Avignon; and his will-power realized this first miracle of his career--he straightway obtained the highest place.

In those days, when education had barely reached the lower classes, the instruction given in the primary normal school was still of the most summary. Spelling, arithmetic, and geometry practically exhausted its resources. As for natural history, a poor despised science, almost unknown, no one dreamed of it, and no one learned or taught it; the syllabus ignored it, because it led to nothing. For Fabre only, notwithstanding, it was his fixed idea, his constant preoccupation, and "while the dictation class was busy around him, he would examine, in the secrecy of his desk, the sting of a wasp or the fruit of the oleander," and intoxicate himself with poetry. (1/9.) His pedagogic studies suffered thereby, and the first part of his stay at the normal school was by no means extremely brilliant. In the middle of his second year he was declared idle, and even marked as an insufficient pupil and of mediocre intelligence. Stung to the quick, he begged as a favour that he should be given the opportunity of following the third year's course in the six months that remained, and he made such an effort that at the end of the year he victoriously won his superior certificate. (1/10.)

A year in advance of the regulation studies, his curiosity might now exercise itself freely in every direction, and little by little it became universal. A chance chemistry lesson finally awakened in him the appetite for knowledge, the passion for all the sciences, of which he thirsted to know at least the elements. Between whiles he returned to his Latin, translating Horace and re-reading Virgil. One day his director put an "Imitation" into his hands, with double columns in Greek and Latin. The latter, which he knew fairly well, assisted him to decipher the Greek. He hastened to commit to memory the vocables, and idioms and phrases of all kinds (1/11.), and in this curious fashion he learned the language. This was his only method of learning languages. It is the process which he recommended to his brother, who was commencing Latin:

"Take Virgil, a dictionary, and a grammar, and translate from Latin into French for ever and for ever; to make a good version you need only common sense and very little grammatical knowledge or other pedantic accessories.

"Imagine an old inscription half-effaced: correctness of judgment partly supplies the missing words, and the sense appears as if the whole were legible. Latin, for you, is the old inscription; the root of the word alone is legible: the veil of an unknown language hides the value of the termination: you have only the half of the words; but you have common sense too, and you will make use of it." (1/12.)

CHAPTER 2.

THE PRIMARY TEACHER.

Furnished with his superior diploma, he left the normal school at the age of nineteen, and commenced as a primary teacher in the College of Carpentras.

The salary of the school teacher, in the year 1842, did not exceed 28 pounds sterling a year, and this ungrateful calling barely fed him, save on "chickpeas and a little wine." But we must beware lest, in view of the increasing and excessive dearness of living in France, the beggarly salaries of the poor schoolmasters of a former day, so little worthy of their labours and their social utility, appear even more disproportionately small than they actually were. What is more to the point, the teachers had no pension to hope for. They could only count on a perpetuity of labour, and when sickness or infirmity arrived, when old age surprised them, after fifty or sixty years of a narrow and precarious existence, it was not merely poverty that awaited them; for many there was nothing but the blackest destitution. A little later, when they began to entertain a vague hope of deliverance, the retiring pension which was held up to their gaze, in the distant future, was at first no more than forty francs, and they had to await the advent of Duruy, the great minister and liberator, before primary instruction was in some degree raised from this ignominious level of abasement.

It was a melancholy place, this college, "where life had something cloistral about it: each master occupied two cells, for, in consideration of a modest payment, the majority were lodged in the establishment, and ate in common at the principal's table."

It was a laborious life, full of distasteful and repugnant duties. We can readily imagine, with the aid of the striking picture which Fabre has drawn for us, what life was in these surroundings, and what the teaching was: "Between four high walls I see the court, a sort of bear-pit where the scholars quarrelled for the space beneath the boughs of a plane-tree; all around opened the class-rooms, oozing with damp and melancholy, like so many wild beasts' cages, deficient in light and air...for seats, a plank fixed to the wall...in the middle a chair, the rushes of the seat departed, a blackboard, and a stick of chalk." (2/1.)

Let the teachers of our spacious and well-lighted schools of to-day ponder on these not so distant years, and measure the progress accomplished. Evoking the memory of their humble colleague of Carpentras, may they feel the true greatness of his example: a noble and a glorious example, of which they may well be proud.

CHAPTER 2.

And what pupils! "Dirty, unmannerly: fifty young scoundrels, children or big lads, with whom," no doubt, "he used to squabble," but whom, after all, he contrived to manage, and by whom he was listened to and respected: for he knew precisely what to say to them, and how, while talking lightly, to teach them the most serious things. For the joy of teaching, and of continually learning by teaching others, made everything endurable. Not only did he teach them to read, write, and cipher, which then included almost the entire programme of primary education; he endeavoured also to place his own knowledge at their service, as he himself acquired it.

It was not only his love of the work that sustained him; it was the desire to escape from the rut, to accomplish yet another stage; to emerge, in short, from so unsatisfactory a position. Now nothing but physical and mathematical science would allow him to entertain the hope of "making an opening" in the world of secondary schoolmasters. He accordingly began to study physics, quite alone, "with an impossible laboratory, experimenting after his own fashion"; and it was by teaching them to his pupils that he learned first of all chemistry, inexpensively performing little elementary experiments before them, "with pipe-bowls for crucibles and aniseed flasks for retorts," and finally algebra, of which he knew not a word before he gave his first lesson. (2/2.)

CHAPTER 2.

How he studied, what was the secret of his method, he told his brother a few years later, when the latter, marking time behind him, was pursuing the same career. A very disappointing career, no doubt, and far from lucrative, but "one of the noblest; one of those best fitted for a noble spirit, and a lover of the good." (2/3.)

Listen to the lesson which he gives his brother:

"To-day is Thursday; nothing calls you out of doors; you choose a thoroughly quiet retreat, where the light is not too strong. There you are, elbows on table, your thumbs to your ears, and a book in front of you. The intelligence awakes; the will holds the reins of it; the outer world disappears, the ear no longer hears, the eye no longer sees, the body no longer exists; the mind schools itself, recollects itself; it is finding knowledge, and its insight increases. Then the hours pass quickly, quickly; time has no measure. Now it is evening. What a day, great God! But hosts of truths are grouped in the memory; the difficulties which checked you yesterday have fused in the fire of reflection; volumes have been devoured, and you are content with your day...

"When something embarrasses you do not abuse the help of your colleagues; with assistance the difficulty is only evaded; with patience and reflection IT IS OVERTHROWN.

Moreover, one knows thoroughly only what one learns oneself; and I advise you earnestly, as far as possible, to have recourse to no aid other than reflection, above all for the sciences. A book of science is an enigma to be deciphered; if some one gives you the key of the enigma nothing appears more simple and more natural than the explanation, but if a second enigma presents itself you will be as unskilful as you were with the first...

"It is probable that you will get the chance of a few lessons; do not by preference accept the easier and more lucrative, but rather the more difficult, even when the subject is one of which as yet you know nothing. The self-esteem which will not allow one's true character to be seen is a powerful aid to the will. Do not forget the method of Jules Janin, running from house to house in Paris for a few wretched lessons in Latin: 'Unable to get anything out of my stupid pupils, with the besotted son of the marquis I was simultaneously pupil and professor: I explained the ancient authors to myself, and so, in a few months, I went through an excellent course of rhetoric...'

"Above all you must not be discouraged; time is nothing provided the will is always alert, always active, and never distracted; 'strength will come as you travel.'

CHAPTER 2. 54

"Try only for a few days this method of working, in which the whole energy, concentrated on one point, explodes like a mine and shatters obstacles; try for a few days the force of patience, strength, and perseverance; and you will see that nothing is impossible!" (2/4.)

These serious reflections show very clearly that his mind was already as mature, as earnest, and as concentrated as it was ever to be.

Not only did he join example to precept; he looked about him and began to observe nature in her own house. The doings of the Mason-bee, which he encountered for the first time, aroused his interest to such a pitch that, being no longer able to constrain his curiosity, he bought--at the cost of what privations!--Blanchard's "Natural History of the Articulata," then a classic work, which he was to re-read a hundred times, and which he still retains, giving it the first place in his modest library, in memory of his early joys and emotions.

The rocks also arrested and captivated his attention: and already the first volumes were corpulent of what was eventually to become his gigantic herbiary. His brother, about to leave for Vezins on vacation, was told of the specimens which he wanted to complete his collection; for although he had never set foot there since his first

CHAPTER 2.

departure, he recalled, with remarkable precision, all the plants that grew in his native countryside; their haunts, their singularities, and the characteristics by which one could not fail to recognize them: as well as all the places which they chose by preference, where he used to wander as an urchin; the Parnassia palustris, "which springs up in the damp meadows, below the beech-wood to the west of the village; which bears a superb white flower at the top of a slightly twisted stem, having an oval leaf about its middle"; the purple digitalis, "whose long spindles of great red flowers, speckled with white inside, and shaped like the fingers of a glove," border a certain road; all the ferns that grow on the wastes, "amid which it is often no easy task to recollect one's whereabouts," and on the arid hills all the heathers, pink, white, and bluish, with different foliage, "of which the innumerable species do not, however, very greatly differ." Nothing is to be neglected; "every plant, whatever it may be, great or little, rare or common, were it only a frond of moss, may have its interest." (2/5.)

Never weary of work, he accumulated all these treasures in his little museum, in order to study them the better; he collected all the coins exhumed from this ancient soil, formerly Roman, "records of humanity more eloquent than books," and which revealed to him the only method of learning and actually re-living history: for he saw in knowledge not merely a means of gaining his bread, but "something nobler; the means of raising the spirit in the contemplation of the truth, of isolating it at will from the miseries of reality, so to find, in these intellectual regions, the only hours of happiness that we may be permitted to taste." (2/6.)

Fabre was so steeped in this passion for knowledge that he wished to evoke it in his brother, now teacher at Lapalud, on the Rhône, not far from Orange. It seemed to him that he would delight in his wealth still better could he share it with another. (2/7.) He stimulated him, pricked him on, and sought to encourage the remarkable aptitude for mathematics with which he believed him endowed. He employed his whole strength in breathing into the other's mind "that taste for the true and the beautiful" which possessed his own nature; he wished to share with him those stores of learning "which he had for some years so painfully amassed"; he would profit by the vacation to place them at his disposal; they would work together "and the light would come." Above all his brother must not allow his intelligence to slumber, must beware of "extinguishing that divine light without which one can, it is true, attend to one's business, but which alone can make a man honourable and respected."

Let him, on the contrary, cultivate his mind incessantly, "the only patrimony on which either of us can count"; the reward would be his moral well-being, and, he hoped, his physical welfare also.

Once more he reinforced his advice by that excellent counsel which was always his own lodestar:

"Science, Frédéric, knowledge is everything...You are too good a thinker not to say with me that no one can better employ his time than by acquiring fresh knowledge...Work, then, when you have the opportunity...an opportunity that very few may possess, and for which you ought to be only too thankful. But I will stop, for I feel my enthusiasm is going to my head, and my reasons are so good already that I have no need of still more triumphant reasons to convince you." (2/8.)

He had only one passion: shooting; more especially the shooting of larks. This sport delighted him, "with the mirror darting its intermittent beams under the rays of the morning sun amid the general scintillation of the dewdrops and crystals of hoarfrost hanging on every blade of grass." (2/9.)

His sight was admirably sure, and he rarely missed his aim. His passion for shooting was always sustained by the same motive: the desire to acquire fresh knowledge; to examine unknown creatures close at hand; to discover

CHAPTER 2.

what they ate and how they lived.

Later, when he again took up his gun, it was still because of his love of life: it was to enable him to enumerate, inventory, and interrogate his new compatriots, his feathered fellow-citizens of Sérignan; to inform himself of their diet, to reveal the contents of their crops and gizzards.

At one time he suddenly ceased to employ this distraction; he seems to have sacrificed it easily, under the stress of present necessities and cruel anxieties as to his uncertain future. "When we do not know where we shall be tomorrow nothing can distract us." (2/10.)

His responsibilities were increasing. He had lately married. On the 30th October, 1844, he was wedded to a young girl of Carpentras, Marie Villard, and already a child was born. His parents, always unlucky, met nowhere with any success. By dint of many wanderings they had finally become stranded at Pierrelatte, the chief town of the canton of La Drôme, sheltered by the great rock which has given the place its name; and there again, of course, they kept a café, situated on the Place d'Armes.

The whole family was now assembled in the same district, a few miles only one from another: but Henri was really its head. Having heard that a quarrel had arisen between his brother and his mother, he wrote to Frédéric in reprimand; gently scolding him and begging him to set matters right, "even if all the wrongs were not on his side."

"My father, in one of his letters, complains that in spite of your nearness you have not yet been to see them. I know very well there is some reason for sulking; but what matter? Give it up: forget everything; do your best to put an end to all these petty and ugly estrangements. You will do so, won't you? I count on it, for the happiness of all." (2/11.)

He was their arbitrator, their adviser, their oracle, their bond of union.

With all this, he was ready to attempt the two examinations which were to decide his future. Very shortly, at Montpellier, he passed almost successively, at an interval of only a few months the examinations for both his baccalauréats; and then the two licentiate examinations in mathematics and physical science.

While he was ardently studying for these examinations, sorrow for the first time knocked at his door. His first-born fell suddenly ill, and in a few days died. On this occasion all his ardent spirituality asserted itself, though in stricken accents, in the letter which he wrote to his brother to announce his loss:

CHAPTER 2.

"After a few days of a marked improvement, which made me think he was saved, two large teeth were cut...and in three days a dreadful fever took him, not from us, who will follow him, but from this miserable world. Ah, poor child, I shall always see you as you were during those last moments, turning those wide, wandering eyes toward heaven, seeking the way to your new country. With a heart full of tears, I shall often let my thoughts go straying after you; but alas! with the eyes of the body I shall never see you again. I shall see you no more: yet only a few days ago I was making the finest plans for you. I used to work for you only; in my studies I thought only of you. Grow up, I used to say, and I will pour into your mind all the knowledge which has cost me so dear, which I am hoarding little by little...But reflection leads me to higher thoughts. I choke back the tears in my heart, and I congratulate him that Heaven has mercifully spared him this life of trials...My poor child...you will never, like your father, have to struggle against poverty and misfortune; you will never know the bitterness of life, and the difficulties of creating a position at a time when there are so many paths that lead to failure...I weep for you because we have lost you, but I rejoice because you are happy...You are happy, and this is not the mad hope of a father broken by sorrow; no, your last glance told me so, too eloquently for me to doubt it. Oh, how beautiful you were in your mortal pallor; the last sigh on your lips, your gaze upon heaven,

CHAPTER 2.

and your soul ready to fly into the bosom of God! Your last day was the most beautiful!" (2/12.)

Although study was his refuge, although he was thereby able to live through these evil days without too greatly feeling their weight, his position was hateful, and he lived a wretched life "from one day to another, like a beggar."

In those troublous times, when education was of no account, it often happened that his teacher's salary was several months in arrears, and the city of Carpentras, "not being in funds," paid it only by instalments, and even so kept him a long time waiting. "One has to besiege the paymaster's door merely to obtain a trifle on account. I am ashamed of the whole business, and I would gladly abandon my claim if I knew where to raise any money." (2/13.)

The genius of Balzac has recorded some unforgettable types of those poor and notable lives, at once so humble and so lofty. He has described the village curé and the country doctor. But how we should have loved to encounter in his gallery, among so many living portraits, a picture of the university life of fifty years ago; and above all a picture of the small schoolmaster of other days, living a life so narrow, so slavish, so painful, and yet so full of worth, so imbued with the sense of duty, and withal so resigned; a portrait for which Fabre might have served as model and prototype, and for which he himself has drawn an unforgettable sketch.

He awaited impatiently the news of his removal, very modestly limiting his ambitions to the hope of entering some lycée as professor of the sciences. His rector was not unnaturally astonished that a young man of such unusual worth, already twice a licentiate, should be so little appreciated by those in high places and allowed to stagnate so long in an inferior post, and one unworthy of him.

In the end, however, after much patient waiting, he became indignant; as always, he could see nothing ahead. The chair of mathematics at Tournon escaped him. Another position, at Avignon, also "slipped through his fingers"; why or how he never knew. He "began to see clearly what life is, and how difficult it is to make one's mark amid all this army of schemers, beggars and imbeciles who besiege every vacant post."

But his heart was "none the less hot with indignation"; he had had enough of "Carpentras, that accursed little hole"; and when the vacations came round once more he "plainly considered the question" and declared "that he would never again set foot inside a communal school." (2/14.)

CHAPTER 2.

He wrote to the rector: "If instead of crushing me into the narrow round of a primary school they would give me some employment of the kind for which my studies and ideas fit me, they would know then what is hatching in my head and what untirable activity there is in me." (2/15.)

He resigned himself nevertheless; he cursed and swore and stormed at his fate; but he had once more to put up with it "for want of a better." All the same "the injustice was too unheard-of, and no one had ever seen or would ever see the like: to give him two licentiate's diplomas, and to make him conjugate verbs for a pack of brats! It was too much!" (2/16.)

CHAPTER 3.

CORSICA.

At last the chair of physics fell vacant at the college of Ajaccio, the salary being 72 pounds sterling, and he left for Corsica. His stay there was well calculated to impress him. There the intense impressionability which the little peasant of Aveyron received at birth could only be confirmed and increased. He felt that this superb and luxuriant nature was made for him, and that he was born for it; to understand and interpret it. He would lose himself in a delicious intoxication, amid the deep woodlands, the mountains rich with scented flowers, wandering through the maquis, the myrtle scrub, through jungles of lentisk and arbutus; barely containing his emotion when he passed beneath the great secular chestnut-trees of Bastelica, with their enormous trunks and leafy boughs, whose sombre majesty inspired in him a sort of melancholy at once poetic and religious. Before the sea, with its infinite distances, he lingered in ecstasy, listening to the song of the waves, and gathering the marvellous shells which the snow-white breakers left upon the beach, and whose unfamiliar forms filled him with delight.

He was soon so accustomed to his new life in peaceful Ajaccio, whose surroundings, decked in eternal verdure, are so captivating and so beautiful, that in spite of a vague desire for change he now dreaded to leave it. He never wearied of admiring and exalting the beautiful and majestic aspects of his new home. How he longed to share his enthusiasm with his father or his brother, as he rambled through the neighbouring maquis!

"The infinite, glittering sea at my feet, the dreadful masses of granite overhead, the white, dainty town seated beside the water, the endless jungles of myrtle, which yield intoxicating perfumes, the wastes of brushwood which the ploughshare has never turned, which cover the mountains from base to summit; the fishing-boats that plough the gulf: all this forms a prospect so magnificent, so striking, that whosoever has beheld it must always long to see it again." (3/1.)

"What is their rock of Pierrelatte, that enormous block of stone which overhangs the place where they dwell, a reef which rises from the surface of the ancient sea of alluvium, compared with these blocks of uprooted granite which lie upon the hillsides here?"

And what were the Aubrac hills which traversed his native country; what was the Ventoux even, that famous Alp, "beside the peaks which rise about the gulf of Ajaccio, always crowned with clouds and whitened with snow, even when the soil of the plains is scorching and rings like a fired brick?"

Time did nothing to abate these first impressions, and after more than a year on the island he was still full of wonder "at the sight of these granite crests, corroded by the severities of the climate, jagged, overthrown by the lightning, shattered by the slow but sure action of the snows, and these vertiginous gulfs through which the four winds of heaven go roaring; these vast inclined planes on which snow-drifts form thirty, sixty, and ninety feet in depth, and across which flow winding watercourses which go to fill, drop by drop, the yawning craters, there to form lakes, black as ink when seen in the shadow, but blue as heaven in the light...

"But it would be impossible for me to give you the least idea of this dizzy spectacle, this chaos of rocks, heaped in frightful disorder. When, closing my eyes, I contemplate these results of the convulsion of the soil in my mind's eye, when I hear the screaming of the eagles, which go wheeling through the bottomless abysses, whose inky shadows the eye dares hardly plumb, vertigo seizes me, and I open my eyes to reassure myself by the reality."

And he sends with his letter a few leaves of the snow immortelle--the edelweiss--plucked on the highest summits, amid the eternal snows; "you will put this in some book, and when, as you turn the leaves, the immortelle meets your eyes, it will give you an excuse for dreaming of the beautiful horrors of its native place." (3/2.)

What a misfortune for him, what regret he would feel, "if he had now to go to some trivial country of plains, where he would die of boredom!"

For him everything was unfamiliar: not only the flora, but the maritime wealth of this singular country. He would set out of a morning, visiting the coves and creeks, roving along the beaches of this magnificent gulf, a lump of bread in his pocket, quenching his thirst with sea-water in default of fresh!

They were mornings full of rosy illusions, whose smiling hopes were revealed in his admirable letters to his brother. Already he meditated a conchology of Corsica, a colossal history of all the molluscs which live upon its soil or in its waters. (3/3.) He collected all the shells he could procure. He analysed, described, classed, and co-ordinated not only the marine species, but the terrestrial and freshwater shells also, extant or fossil. He asked his brother to collect for him all the shells he could find in the marshes of Lapalud, in the brooks and ditches of the neighbourhood of Orange. In his enthusiasm he tried to convince him of the immense interest of these researches, which might perhaps seem ridiculous or futile to him; but let him only think of geology; the humblest shell picked up might throw a sudden light upon the formation of this or that stratum. None are to be disdained: for men have considered, with reason, that they were honouring the memory of their eminent fellows by giving their names to the rarest and most beautiful. Witness the magnificent Helix dedicated to Raspail, which is found only in the caverns where the strawberry-tree grows amid the high mountains of Corsica. (3/4.)

Moreover, he said, "the infinitesimal calculus of Leibnitz will show you that the architecture of the Louvre is less learned than that of a snail: the eternal geometer has unrolled his transcendent spirals on the shell of the mollusc that you, like the vulgar profane, know only seasoned with spinach and Dutch cheese." (3/5.)

For all that, he did not neglect his mathematics, in which, on the contrary, he found abundant and suggestive recreation. The properties of a figure or a curve which he had newly discovered prevented his sleep for several nights.

"All this morning I have been busy with star-shaped polygons, and have proceeded from surprise to surprise...perceiving in the distance, as I advanced, unforeseen and marvellous consequences."

Here, among others, is one question which suddenly presented itself to his mind "in the midst of the spikes" of his polygons: what would be the period of the rotation of the sun on its own centre if its atmosphere reached as far as the earth? And this question gave rise to another, "without which the sequence stops then and there; number, space, movement, and order form a single chain, the first link of which sets all the rest in motion." (3/6.) And the hours went by quickly, so quickly with "x," the plants and the shells, that "literally there was no time to eat."

For Fabre was born a poet, and mathematics borders upon poetry; he saw in algebra "the most magnificent flights," and the figures of analytical geometry unrolled themselves in his imagination "in superb strophes"; the Ellipse, "the trajectory of the planets, with its two related foci, sending from one to the other a constant sum of vector radii"; the Hyperbole, "with repulsive foci, the desperate curve which plunges into space in infinite tentacles, approaching closer and closer to a straight line, the asymptote, without ever finally attaining it"; the Parabola, "which seeks fruitlessly in the infinite for its second, lost centre: it is the trajectory of the bomb: it is the path of certain comets which come one day to visit our sun, then flee into the depths whence they never return." (3/7.)

And one fine morning we behold him mounting, thrilled by a lyric passion, to the lofty regions in which Number, "irresistible, omnipotent, keystone of the vault of the universe, rules at once Time and Space." He ascends, he rushes forward, farther than the chariot--

"Beyond the Husbandman who ploughs in space And sows the suns in furrows of the skies."

He ascends those tracks of flame, where on high

"in those lists inane Wise regulator, Number holds the reins Of those indomitable steeds; Number has set a bit i' the foaming mouths Of these Leviathans, and with nervous hand Controls them in their tracks;

Their smoking flanks beneath the yoke in vain Quiver; their nostrils vainly void as foam Dense tides of lava; and in vain they rear; For Number on their mettled haunches poised Holds them, or duly with the rein controls, Or in their flanks buries his spur divine." (3/8.)

Later he confessed all that he owed, as a writer, to geometry, whose severe discipline forms and exercises the mind, gives it the salutary habit of precision and lucidity, and puts it on its guard against terms which are incorrect or unduly vague, giving it qualities far superior to all the "tropes of rhetoric."

It was then that he became the pupil of Requien of Avignon, the retired botanist, a lofty but somewhat limited mind, who was hardly capable of opening up other horizons to him. But Requien did at least enrich his memory by a prodigious quantity of names of plants with which he had not been acquainted. He revealed to him the immense flora of Corsica, which he himself had come to study, and for which Fabre was to gather such a vast amount of material.

Fabre found in Requien more especially a friend "proof against anything"; and when the latter died almost suddenly at Bonifacio, Fabre was overwhelmed by the sad news. On that very day he had on the table before him a parcel of plants gathered for the dead botanist. "I cannot let my eyes rest upon it," he wrote at the time, "without feeling my heart wrung and my sight dim with tears." (3/9.)

But the most admirably fruitful encounter, as it exercised the profoundest influence upon his destiny, was his meeting with Moquin-Tandon, a Toulouse professor who followed Requien to Corsica, to complete the work which the latter had left unfinished: the complete inventory of the prodigious wealth of vegetation, of the innumerable species and varieties which Fabre and he collected together, on the slopes and summits of Monte Renoso, often botanizing "up in the clouds, mantle on back and numb with cold." (3/10.)

Moquin-Tandon was not merely a skilful naturalist; he was one of the most eloquent and scholarly scientists of his time. Fabre owed to him, not his genius, to be sure, but the definite indication of the path he was finally to take, and from which he was never again to stray.

Moquin-Tandon, a brilliant writer and "an ingenious poet in his Montpellerian dialect," (3/11.) taught Fabre never to forget the value of style and the importance of form, even in the exposition of a purely descriptive science such as botany. He did even more, by one day suddenly showing Fabre, between the fruit and the cheese, "in a plate of water," the anatomy of the snail. This was his first introduction to his true destiny before the final revelation of which I shall presently speak. Fabre understood then and there that he could do decidedly better than to stick to mathematics, though his whole career would feel the effects of that study.

"Geometers are made; naturalists are born ready-made," he wrote to his brother, still excited by this incident, "and you know better than any one whether natural history is not my favourite science." (3/12.)

CHAPTER 3. 73

>From that time forward he began to collect not only dead, inert, or dessicated forms, mere material for study, with the aim of satisfying his curiosity; he began to dissect with ardour, a thing he had never done before. He housed his tiny guests in his cupboard; and occupied himself, as he was always to do in the future, with the smaller living creatures only.

"I am dissecting the infinitely little; my scalpels are tiny daggers which I make myself out of fine needles; my marble slab is the bottom of a saucer; my prisoners are lodged by the dozen in old match-boxes; maxime miranda in minimis." (3/13.)

Roaming at night along the marshy beaches, he contracted fever, and several terrible attacks, accompanied by alarming tremors, left him so bloodless and feeble that, much against his will, he had to beg for relief, and even insist upon his prompt return to the mainland. in the meantime he obtained sick-leave, and returned to Provence after a terrible crossing which lasted no less than three days and two nights, on a sea so furious that he gave himself up for lost. (3/14.)

Slowly he recovered his health, and after a second but brief stay at Ajaccio he received the news of his appointment to the lycée of Avignon. (3/15.)

CHAPTER 3.

He returned with his imagination enriched and his mind expanded, with settled ideas, and thoroughly ripe for his task.

CHAPTER 4.

AT AVIGNON.

The resolute worker resumed his indefatigable labours with an ardour greater than ever, for now he was haunted by a noble ambition, that of becoming a teacher of the superior grade, and of "talking plants and animals" in a chair of the faculty. With this end in view he added to his two diplomas--those of mathematics and physics--a third certificate, that of natural sciences. His success was triumphant.

Already tenacious and fearless in affirming what he believed to be the truth, he astonished and bewildered the professors of Toulouse. Among the subjects touched upon by the examiners was the famous question of spontaneous generation, which was then so vital, and which gave rise to so many impassioned discussions. The examiner, as it chanced, was one of the leading apostles of this doctrine. The future adversary of Darwin, at the risk of failure, did not scruple to argue with him, and to put forward his personal convictions and his own arguments. He decided the vexed question in his own way, on his own responsibility. A personality already so striking was regarded with admiration; a candidate so far out of the ordinary was welcomed with enthusiasm, and but for the insufficiency of the budget which so scantily met the needs of public instruction his examination fees would have been returned. (4/1.)

Why, after this brilliant success, was Fabre not tempted to enter himself for a fellowship, which would later in his career have averted so many disappointments? It was doubtless because he felt, obscurely, that his ideal future lay along other lines, and that he would have been taking a wrong turning. Despite all the solicitations which were addressed to him he would think of nothing but "his beloved studies in natural history" (4/2.); he feared to lose precious time in preparing himself for a competitive examination; "to compromise by such labour, which he felt would be fruitless" (4/3.), the studies which he had already commenced, and the inquiries already carried out in Corsica. He was busy with his first original labours, the theses which he was preparing with a view to his doctorate in natural science, "which might one day open the doors of a faculty for him, far more easily than would a fellowship and its mathematics." (4/4.)

At heart he was utterly careless of dignities and degrees. He worked only to learn, not to attain and follow up a settled calling. What he hoped above all was to succeed in devoting all his leisure to those marvellous natural sciences in which he could vaguely foresee studies full of interest; something animated and vital; a thousand fascinating themes, and an atmosphere of poetry.

His genius, as yet invisible, was ripening in obscurity, but was ready to come forth; he lacked only the propitious circumstance which would allow him to unfold his wings.

He was seeking them in vain when a volume by Léon Dufour, the famous entomologist, who then lived in the depths of the Landes, fell by chance into his hands, and lit the first spark of that beacon which was presently to decide the definite trend of his ideas.

It was this incident which then and there developed the germs already latent within him. These had only awaited such an occasion as that which so fortunately came to pass one evening of the winter of 1854.

Fabre offers yet another example of the part so often played by chance in the manifestations of talent. How many have suddenly felt the unexpected awakening of gifts which they did not suspect, as a result of some unusual circumstance!

Was it not simply as a result of having read a note by the Russian chemist Mitscherlich on the comparison of the specific characteristics of certain crystals that Pasteur so enthusiastically took up his researches into molecular asymmetry which were the starting-point of so many wonderful discoveries?

Again, we need only recall the case of Brother Huber, the celebrated observer of the bee, who, having out of simple curiosity undertaken to verify certain experiments of Réaumur's, was so completely and immediately fascinated by the subject that it became the object of the rest of his life.

Again, we may ask what Claude Bernard would have been had he not met Magendie? Similarly Léon Dufour's little work was to Fabre the road to Damascus, the electric impulse which decided his vocation.

It dealt with a very singular fact concerning the manners of one of the hymenoptera, a wasp, a Cerceris, in whose nest Dufour had found small coleoptera of the genus Buprestis, which, under all the appearances of death, retained intact for an incredible time their sumptuous costume, gleaming with gold, copper, and emerald, while the tissues remained perfectly fresh. In a word, the victims of Cerceris, far from being desiccated or putrefied, were found in a state of integrity which was altogether paradoxical.

CHAPTER 4.

Dufour merely believed that the Buprestes were dead, and he gave an attempted explanation of the phenomenon.

Fabre, his curiosity and interest aroused, wished to observe the facts for himself; and, to his great surprise, he discovered how incomplete and insufficiently verified were the observations of the man who was at that time known as "the patriarch of entomologists."

>From that moment he saw his way ahead; he suspected that there was still much to discover and much to revise in this vast department of nature, and conceived the idea of resuming the work so splendidly outlined by Réaumur and the two Hubers, but almost completely neglected since the days of those illustrious masters. He divined that here were fresh pastures, a vast unexplored country to be opened up, an entire unimagined science to be founded, wonderful secrets to be discovered, magnificent problems to be solved, and he dreamed of consecrating himself unreservedly, of employing his whole life in the pursuit of this object; that long life whose fruitful activity was to extend over nearly ninety years, and which was to be so "representative" by the dignity of the man, the probity of the expert, the genius of the observer, and the originality of the writer.

CHAPTER 4.

The year 1855 saw the first appearance, in the "Annales des sciences naturelles," of the famous memoir which marked the beginning of his fame: the history, which might well be called marvellous and incredible, of the great Cerceris, a giant wasp and "the finest of the Hymenoptera which hunt for booty at the foot of Mont Ventoux." (4/5.)

Fabre was now thirty-two years old, and his situation as assistant- professor of physics was somewhat precarious. From the 72 pounds sterling which he drew at Ajaccio, an overseas post, his salary was reduced, on his return to the mainland, to 64 pounds sterling, and during the whole of his stay at Avignon he obtained neither promotion nor the smallest increase of pay, excepting a few additional profits which were unconnected with his habitual duties. When he left the university after twenty well-filled years, he left as he had entered, with the same title, rank, and salary of a mere assistant-professor.

Yet all about him "everywhere and for every one, all was black indeed": his family had increased and therewith his expenses; there were now seven at table every day. Very shortly his modest salary would no longer suffice; he was obliged to supplement it by all sorts of hack-work--classes, "repetitions," private lessons; tasks which repelled him, for they absorbed all his available time; they prevented him from giving himself up to his favourite studies, to his silent and solitary observations. Nevertheless, he acquitted himself of these duties patiently and conscientiously, for at heart he loved his profession, and was rather a fellow-disciple than a master to his pupils. For this reason all those about him worked with praiseworthy assiduity; even the worst elements, the black sheep, the "bad eggs" of other classes, with him were suddenly transformed and as attentive as the rest. Although he knew how to keep order, how to make himself respected, and could on occasion deal severely and speak sternly, so that very few dared to forget themselves before him, he knew also how to be merry with his pupils, chatting with them familiarly, putting himself in their place, entering into their ideas, and making himself their rival. If life was laborious under his ferula, it was also merry. The best proof of this is the fact that of all his colleagues at the lycée he was the only one who had no nickname, a rarity in scholastic annals.

He did not therefore object to these lessons; but while at Carpentras he was made much of and praised by the principal, was a general favourite, and had perfect liberty to follow his inspiration during his partly gratuitous classes, here the hours and the programme tied him down, which was precisely what he found insupportable.

Everything made things difficult for him here: his external self; his character, ever so little shy and unsocial; his temperament, which was made for solitude.

In the thick of this hierarchical society of university professors he remained independent; he knew nothing of what was said or what was happening in the college, and his colleagues were always better informed than he. (4/6.) As he was not a fellow, he was made to feel the fact and was treated as a subordinate; the others, who prided themselves on the title, and who were incapable of recognizing his merit, which was a little beyond them, were jealous of him, all the more inasmuch as his name was momentarily noised abroad, and they revenged themselves by calling him "the fly" among themselves, by way of allusion to his favourite subject. (4/7.)

Indifferent to distinctions, as well as to those who bore them, contemptuous of etiquette, and incapable of putting constraint upon his nature, he remained an "outsider," and refused to comply with a host of factitious or worldly obligations which he regarded as useless or disgusting. Thus even at Ajaccio he managed to escape the customary ceremonies of New Year's Day.

"Good society I avoid as much as possible; I prefer my own company. So I have seen no one; I did not respond to the principal's invitation to make the official round of visits." (4/8.)

CHAPTER 4.

When obliged to accept some invitation, apart from occasions of too great solemnity, when he was really constrained to dress himself in the complete livery of circumstance and ceremony, he remained faithful to his black felt hat, which made a blot among all the carefully polished "toppers" of his colleagues. He was called to order; he was reprimanded; he obeyed unwillingly, or worse, he resisted; he revolted, and threatened to send in his resignation. To pay court to people, to endeavour to make himself pleasant, to grovel before a superior, were to him impossibilities. He could neither solicit, nor sail with the wind, nor force himself on others, nor even make use of his relations.

However, when he went to Paris to take his doctor's degree in natural sciences, he did not forget Moquin-Tandon, who had formerly, in Corsica, revealed to him the nature of biology, and whom he himself had received and entertained in his humble home.

The ex-professor of Toulouse, who was now eminent in his speciality, occupied the chair of natural history in the faculty of medicine in Paris. What better occasion could he wish of introducing himself to a highly placed official? Fabre had formerly been his host; he could recall the happy hours they had spent together; he could explain his plans, and ask for the professor's assistance! Fate pointed to him as a protector. But if Fabre had been capable of climbing the professor's stairs with some such ambitious desires, he would quickly have been disabused.

The "dear master" had long ago forgotten the little professor of Ajaccio, and his welcome was by no means such as Fabre had the right to expect. Far from insisting, he was disheartened, perhaps a little humiliated, and hastened to take his leave.

The theses which Fabre brought with him, and which, he had thought, ought to lead him one day to a university professorship, did not, as a matter of fact, contain anything very essentially original.

He had been attracted, indeed fascinated, by all the singularities presented by the strange family of the orchids; the asymmetry of their blossoms, the unusual structure of their pollen, and their innumerable seeds; but as for the curious rounded and duplicated tubercles which many of them bore at their base, what precisely were they? The greatest botanists--de Candolle, A. de Jussieu--had perceived in them nothing more than roots. Fabre demonstrated in his thesis that these singular organs are in reality merely buds, true branches or shoots, modified and disguised, analogous to the metamorphosed tubercle of the potato. (4/9.)

CHAPTER 4.

He added also a curious memoir on the phosphorescence of the agaric of the olive-tree, a phenomenon to which he was to return at a later date.

In the field of zoology his scalpel revealed the complicated structure of the reproductive organs of the Centipedes (Millepedes), hitherto so confused and misunderstood; as also certain peculiarities of the development of these curious creatures, so interesting from the point of view of the zoological philosopher (4/10.), for he had become expert in handling not only the magnifying glass, which was always with him, but also the microscope, which discovers so many infinite wonders in the lowest creatures, yet which was not of particular service in any of the beautiful observations upon which his fame is built.

Returning to Avignon, in the possession of his new degree, he commenced an important task which took him nearly twenty years to complete: a painstaking treatise on the Sphaeriaceae of Vaucluse, that singular family of fungi which cover fallen leaves and dead twigs with their blackish fructifications; a remarkable piece of work, full of the most valuable documentation, as were the theses whose subjects I have just detailed; but without belittling the fame of their author, one may say that another, in his place, might have acquitted himself as well.

CHAPTER 4.

Although he continued to undertake researches of limited interest and importance, although he persisted in dissecting plants, and, although he disliked it, in "disembowelling animals," the fact was that apart from Thursdays and Sundays it was scarcely possible for him to escape from his week's work; hardly possible to snatch sufficient leisure to undertake the studies toward which he felt himself more particularly drawn. Tied down by his duties, which held him bound to a discipline that only left him brief moments, and by the forced hack-work imposed upon him by the necessity of earning his daily bread, he had scarcely any time for observation excepting vacations and holidays.

Then he would hasten to Carpentras, happy to hold the key to the meadows, and wander across country and along the sunken lanes, collecting his beautiful insects, breathing the free air, the scent of the vines and olives, and gazing upon Mont Ventoux, close at hand, whose silver summit would now be hidden in the clouds and now would glitter in the rays of the sun.

Carpentras was not merely the country in which his wife's parents dwelt: it was, above all, a unique and privileged home for insects; not on account of its flora, but because of the soil, a kind of limestone mingled with sand and clay, a soft marl, in which the burrowing hymenoptera could easily

CHAPTER 4.

establish their burrows and their nests. Certain of them, indeed, lived only there, or at least it would have been extremely difficult to find them elsewhere; such was the famous Cerceris; such again, was the yellow-winged Sphex, that other wasp which so artistically stabs and paralyses the cricket, "the brown violinist of the clods."

At Carpentras too the Anthophorae lived in abundance; those wild bees with whom the vexed and enigmatic history of the Sitaris and the Meloë is bound up; those little beetles, cousins of the Cantharides, whose complex metamorphoses and astonishing and peculiar habits have been revealed by Fabre. This memoir marked the second stage of his scientific career, and followed, at an interval of two years, the magnificent observations on the Cerceris.

These two studies, true masterpieces of science, already constituted two excellent titles to fame, and would by themselves have sufficed to fill a naturalist's whole lifetime and to make his name illustrious.

>From that time forward he had no peer. The Institute awarded him one of its Montyon prizes (4/11.), "an honour of which, needless to say, he had never dreamed." (4/12.) Darwin, in his celebrated work on the "Origin of Species," which appeared precisely at this moment, speaks of Fabre somewhere as "the inimitable observer." (4/13.)

CHAPTER 4.

Exploring the immediate surroundings of Avignon, he very soon discovered fresh localities frequented almost exclusively by other insects, whose habits in their turn absorbed his whole attention.

First of these was the sandy plateau of the Angles, where every spring, in the sunlit pastures so beloved of the sheep, the Scarabaeus sacer, with his incurved feet and clumsy legs, commences to roll his everlasting pellet, "to the ancients the image of the world." His history, since the time of the Pharaohs, had been nothing but a tissue of legends; but stripping it of the embroidery of fiction, and referring it to the facts of nature, Fabre demonstrated that the true story is even more marvellous than all the tales of ancient Egypt. He narrated its actual life, the object of its task, and its comical and exhilarating performances. But such is the subtlety of these delicate and difficult researches that nearly forty years were required to complete the study of its habits and to solve the mystery of its cradle. (4/14.)

On the right bank of the Rhône, facing the embouchure of the Durance, is a small wood of oak-trees, the wood of Des Issarts. This again, for many reasons, was one of his favourite spots. There, "lying flat on the ground, his head in the shadow of some rabbit's burrow," or sheltered from the sun by a great umbrella, "while the blue-winged locusts

CHAPTER 4.

frisked for joy," he would follow the rapid and sibilant flight of the elegant Bembex, carrying their daily ration of diptera to her larvae, at the bottom of her burrow, deep in the fine sand." (4/15.)

He did not always go thither alone: sometimes, on Sundays, he would take his pupils with him, to spend a morning in the fields, "at the ineffable festival of the awakening of life in the spring." (4/16.)

Those most dear to him, those who in the subsequent years have remained the object of a special affection, were Devillario, Bordone, and Vayssières (4/17.), "young people with warm hearts and smiling imaginations, overflowing with that springtime sap of life which makes us so expansive and so eager to know.

Among them he was "the eldest, their master, but still more their companion and friend"; lighting in them his own sacred fire, and amazing them by the deftness of his fingers and the acuteness of his lynx-like eyes. Furnished with a notebook and all the tools of the naturalist--lens, net, and little boxes of sawdust steeped in anaesthetic for the capture of rare specimens-- they would wander "along the paths bordered with hawthorn and hyaebla, simple and childlike folk," probing the bushes, scratching up the sand, raising stones, running the net along hedge and meadow, with explosions of delight when they made some splendid capture or discovered some unrecorded marvel of the entomological world.

It was not only on the banks of the Rhône or the sandy plateau of Avignon that they sought adventure thus, "discussing things and other things," but as far as the slopes of Mont Ventoux, for which Fabre had always felt an inexplicable and invincible attraction, and whose ascent he accomplished more than twenty times, so that at last he knew all its secrets, all the gamut of its vegetation, the wealth of the varied flora which climb its flanks from base to summit, and which range "from the scarlet flowers of the pomegranate to the violet of Mont Cenis and the Alpine forget-me-not" (4/18.), as well as the antediluvian fauna revealed amid its entrails, a vast ossuary rich in fossils.

His disciples, all of whom, without exception, regarded him with absolute worship, have retained the memory of his wit, his enthusiasm, his geniality and his infectious gaiety, and also of the singular uncertainty of his temperament; for on some days he would not speak a word from the beginning to the end of his walk.

Even his temper, ordinarily gentle and easy, would suddenly become hasty and violent, and would break out into terrible explosions when a sudden annoyance set him beside himself; for instance, when he was the butt of some ill-natured trick, or when, in spite of the lucidity of his explanations, he felt that he had not been properly understood. Perhaps he inherited this from his mother, a rebellious, crotchety, somewhat fantastic person, by whose temper he himself had suffered.

But the young people who surrounded him were far from being upset by these contrasts of temperament, in which they themselves saw nothing but natural annoyance, and the corollary, as it were, of his abounding vitality. (4/19.)

It was because he was the only university teacher in Avignon to occupy himself with entomology that Pasteur visited him in 1865. The illustrious chemist had been striving to check the plague that was devastating the silkworm nurseries, and as he knew nothing of the subject which he proposed to study, not even understanding the constitution of the cocoon or the evolution of the silkworm, he sought out Fabre in order to obtain from his store of entomological wisdom the elementary ideas which he would find indispensable. Fabre has told us, in a moving page (4/20), with what a total lack of comprehension of "poverty in a black coat" the great scientist gazed at his poor home. Preoccupied by another problem, that of the amelioration of wines by means of heat, Pasteur asked him point-blank-- him, the humble proletarian of the university caste, who drank only the cheapest wine of the country--to show him his cellar. "My cellar! Why not my vaults, my dusty bottles, labelled according to age and vintage! But Pasteur insisted. Then, pointing with my finger, I showed him, in a corner of the kitchen, a chair with all the straw gone, and on this chair a two- gallon demijohn: 'There is my cave, monsieur!'"

If the country professor was embarrassed by the chilliness of the other, he was none the less shocked by his attitude. It would seem, from what Fabre has said, that Pasteur treated him with a hauteur which was slightly disdainful. The ignorant genius questioned his humble colleague, distantly giving him his orders, explaining his plans and his ideas, and informing him in what directions he required assistance.

After this, we cannot be surprised if the naturalist was silent. How could sympathetic relations have survived this first meeting? Fabre could not forgive it. His own character was too independent to accommodate itself to Pasteur's. Yet never, perhaps, were two men made for a better understanding. They were equally expert in exercising their admirable powers of vision in the vast field of nature, equally critical of self, equally careful never to depart from the strict limits of the facts; and they were, one may say, equally eminent in the domain of invention, different though their fortunes may have been; for the sublimity of scientific discoveries, however full of genius they may be, is often measured only by the immediate consequences drawn therefrom and the practical importance of their results.

In reality, were they not two rivals, worthy of being placed side by side in the paradise of sages? Both of them, the one by demolishing the theory of spontaneous generation, the other by refuting the mechanical theory of the origin of instincts, have brought into due prominence the great unknown and mysterious forces which seem destined to hold eternally in suspense the profound enigma of life.

Now he was anxious not to leave the Vaucluse district, the scene of his first success, and a place so fruitful in subjects of study. He wished to remain close to his insects, and also near the precious library and the rich collections which Requien had left by will to the town of Avignon. In spite of the meagreness of his salary, he asked for nothing more; and, what is more, by an inconsequence which is by no means incomprehensible, he avoided everything that might have resulted in a more profitable position elsewhere, and evaded all proposals of further promotion. Twice, at Poitiers and Marseilles, he refused a post as assistant professor, not regarding the advantages sufficient to balance the expenses of removal. (4/21.)

CHAPTER 4.

It is true that his modest position was slightly improved; at the lycée he had just been appointed drawing-master, thanks to his knowledge of design, for he could draw--indeed, what could he not do? The city, on the other hand, appointed him conservator of the Requien Museum, and presently municipal lecturer, so that his earnings were increased by 48 pounds sterling per annum, and he was at last able to abandon "those abominable private lessons" (4/22.), which the insufficiency of his income had hitherto forced him to accept. These new duties, which naturally demanded much time and much labour, kept him almost as badly tied as he had been before.

To be rich enough to set himself free; to be master of all his time, to be able to devote himself entirely to his chosen work: this was his dream, his constant preoccupation: it haunted him; it was a fixed idea.

Such was the principal motive of his inquiry into the properties of madder, the colouring principle of which he succeeded in extracting directly, by a perfectly simple method, which for a time very advantageously replaced the extremely primitive methods of the old dyers, who used a simple extract of madder; a crude preparation which necessitated long and expensive manipulations. (4/23.)

CHAPTER 4.

He had been working at this for eight years when Victor Duruy, Minister of Public Instruction and Grand Master of the University, came to surprise him in his laboratory at Saint-Martial, in the full fever of research. Whatever was Duruy's idea in entering into relations with him, it seems that from their first meeting the two men were really taken with one another: there were, between them, so many close affinities of taste and character. Duruy found in Fabre a man of his own temper; for his, like Fabre's, was a modest and simple nature. Both came of the people, and the principal motive of each was the same ideal of work, emancipation, and progress.

A little later Duruy summoned the modest sage of Avignon to Paris, with particular insistence; he was full of attentions and of forethought, and made him there and then a chevalier of the Legion of Honour; a distinction of which Fabre was far from being proud, and which he was careful never to obtrude; but he nevertheless always thought of it with a certain tenderness, as a beloved "relic" in memory of this illustrious friend.

On the following day the naturalist was conveyed to the Tuileries to be presented to the Emperor. You must not suppose that he was in the least disturbed at the idea of finding himself face to face with royalty. In the presence of all these bedizened folk, in his coat of a cut which was doubtless already superannuated, he cared little for the impression he might produce. As good an observer of men as of beasts, he gazed quietly about him; he exchanged a few words with the Emperor, who was "quite simple," almost suppressed, his eyes always half-closed; he watched the coming and going of "the chamberlains with short breeches and silver- buckled shoes, great scarabaei, clad with café au lait wing-cases, moving with a formal gait." Already he sighed regretfully; he was bored; he was on the rack, and for nothing in the world would he have repeated the experience. He did not even feel the least desire to visit the vaunted collections of the Museum. He longed to return; to find himself once more among his dear insects; to see his grey olive-trees, full of the frolicsome cicadae, his wastes and commons, which smelt so sweet of thyme and cypress; above all, to return to his furnace and retorts, in order to complete his discovery as quickly as possible.

But others profited by his happy conceptions. Like the cicada, the Cigale of his fable (See "Social Life in the Insect World," by Jean-Henri Fabre (T. Fisher Unwin, 1912).), which makes a "honeyed reek" flow from--

"the bark Tender and juicy, of the bough,"

on which it is quickly supplanted by

"Fly, drone, wasp, beetle too with hornèd head" (4/24.),

who

"Now lick their honey'd lips, and feed at leisure,"

so, after he had painfully laboured for twelve years in his well, he saw others, more cunning than he, come to his perch, who by dint of "stamping on his toe," succeeded in ousting him. Pending the appearance of artificial alizarine, which was presently to turn the whole madder industry upside down, these more sophisticated persons were able to benefit at leisure by the ingenious processes discovered by Fabre, so that the practical result of so much assiduity, so much patient research, was absolutely nil, and he found himself as poor as ever.

So faded his dream: and, if we except his domestic griefs, this was certainly the deepest and cruellest disappointment he had ever experienced.

Thenceforth he saw his salvation only in the writing of textbooks, which were at last to throw open the door of freedom. Already he had set to work, under the powerful stimulus of Duruy, preoccupied as he always was by his incessant desire for freedom. The first rudiments of his "Agricultural Chemistry," which sounded so fresh a note in the matter of teaching, had given an instance and a measure of his capabilities.

But he did not seriously devote himself to this project until after the industrial failure and the distressing miscarriage of his madder process; and not until he had been previously assured of the co-operation of Charles Delagrave, a young publisher, whose fortunate intervention contributed in no small degree to his deliverance. Confident in his vast powers of work, and divining his incomparable talent as POPULARIZER, Delagrave felt that he could promise Fabre that he would never leave him without work; and this promise was all the more comforting, in that the University, despite his twenty-eight years of assiduous service, would not accord him the smallest pension.

Victor Duruy was the great restorer of education in France, from elementary and primary education, which should date, from his great ministry, the era of its deliverance, to the secondary education which he himself created in every part. He was also the real initiator of secular instruction in France, and the Third Republic has done little but resume his work, develop his ideas, and extend his programme. Finally, by instituting classes for adults, the evening classes which enabled workmen, peasants, bourgeois, and young women to fill the gaps in their education, he gave reality to the generous and fruitful idea that it is possible for all to divide life into two parts, one having for its object our material needs and our daily bread, and the other consecrated to the spiritual life and the delights of the Ideal.

At the same time he emancipated the young women of France, formerly under the exclusive tutelage of the clergy, and opened to them for the first time the golden gates of knowledge; an audacious innovation, and formidable withal, for it shrewdly touched the interests of the Church, struck a blow at her ever-increasing influence, and clashed with her consecrated privileges and age-long prejudices. (4/25.)

At Avignon Fabre was instructed to give his personal services. He gave them with all his heart; and it was then that he undertook, in the ancient Abbey of Saint-Martial, those famous free lectures which have remained celebrated in the memory of that generation. There, under the ancient Gothic vault, among the pupils of the primary Normal College, an eager crowd of listeners pressed to hear him; and among the most assiduous was Roumanille, the friend of Mistral, he who so exquisitely wove into his harmonies "the laughter of young maidens and the flowers of springtime." No one expounded a fact better than Fabre; no one explained it so fully and so clearly. No one could teach as he did, in a fashion so simple, so animated, so picturesque, and by methods so original.

He was indeed convinced that even in early childhood it was possible for both boys and girls to learn and to love many subjects which had hitherto never been proposed; and in particular that Natural History which to him was a book in which all the world might read, but that university methods had reduced it to a tedious and useless study in which the letter "killed the life."

He knew the secret of communicating his conviction, his profound faith, to his hearers: that sacred fire which animated him, that passion for all the creatures of nature.

These lectures took place in the evening, twice a week, alternately with the municipal lectures, to which Fabre brought no less application and ardour. In the intention of those who instituted them these latter were above all to be practical and scientific, dealing with science applied to agriculture, the arts, and industry.

But might he not also expect auditors of another quality, in love only with the ideal, "who, without troubling about the possible applications of scientific theory, desired above all to be initiated into the action of the forces which rule nature, and thereby to open to their minds more wondrous horizons"?

CHAPTER 4.

Such were the noble scruples which troubled his conscience, and which appeared in the letter which he addressed to the administration of the city, when he was entrusted by the latter with what he regarded as a lofty and most important mission.

"...Is it to be understood that every purely scientific aspect, incapable of immediate application, is to be rigorously banished from these lessons? Is it to be understood that, confined to an impassable circle, the value of every truth must be reckoned at so much per hundred, and that I must silently pass over all that aims only at satisfying a laudable desire of knowledge? No, gentlemen, for then these lectures would lack a very essential thing: the spirit which gives life!" (4/26.)

Physically, according to the testimony of his contemporaries, he was already as an admirable photograph represents him twenty years later: he wore a large black felt hat; his face was shaven, the chin strong and wilful, the eyes vigilant, deep-set and penetrating; he hardly changed, and it was thus I saw him later, at a more advanced age.

The ancient Abbey of Saint-Martial, where these lectures were given, was occupied also by the Requien Museum, of which Fabre had charge. It was here that he one day met

CHAPTER 4.

John Stuart Mill.

The celebrated philosopher and economist had just lost his wife: "the most precious friendship of his life" was ended. (4/27.) It was only after long waiting that he had been able to marry her. Subjected at an early age by a father devoid of tenderness and formidably severe to the harshest of disciplines, he had learned in childhood "what is usually learned only by a man." Scarcely out of his long clothes, he was construing Herodotus and the dialogues of Plato, and the whole of his dreary youth was spent in covering the vast field of the moral and mathematical sciences. His heart, always suppressed, never really expanded until he met Mrs. Harriett Taylor.

This was one of those privileged beings such as seem as a rule to exist only in poetry and literature; a woman as beautiful as she was astonishingly gifted with the rarest faculties; combining with the most searching intelligence and the most persuasive eloquence so exquisite a sensitiveness that she seemed often to divine events in advance.

Mill possessed her at last for a few years only, and he had resigned his post in the offices of the East India Company to enjoy a studious retreat in the enchanted atmosphere of southern Europe when suddenly at Avignon Harriett Mill

CHAPTER 4.

was carried off by a violent illness. (Mill retired in 1858, when the government of India passed to the Crown. He had married Mrs. John Taylor in 1851. [Tr.])

>From that time the philosopher's horizon was suddenly contracted to the limit of those places whence had vanished the adored companion and the beneficent genius who had been the sole charm of his entire existence. Overwhelmed with grief, he acquired a small country house in one of the least frequented parts of the suburbs of Avignon, close to the cemetery where the beloved dead was laid to rest for ever. A silent alley of planes and mulberry-trees led to the threshold, which was shaded by the delicate foliage of a myrtle. All about he had planted a dense hedge of hawthorn, cypress, and arborvitae, above which, from the vantage of a small terrace, built, under his orders, at the level of the first floor, he could see, day by day and at all hours, the white tomb of his wife, and a little ease his grief.

Thus he cloistered himself, "living in memory," having no companion but the daughter of his wife; trying to console himself by work, recapitulating his life, the story of which he has told in his remarkable "Memoirs." (4/28.)

Fabre paid a few visits to this Thebaïd. A solitary such as Mill had become could be attracted only by a man of his temper, in whom he found, if not an affinity of nature, at least tastes like his own, and immense learning, as great as his. For Mill also was versed in all the branches of human knowledge: not only had he meditated on the high problems of history and political economy, but he had also probed all branches of science: mathematics, physics, and natural history. It was above all botany which served them as a bond of union, and they were often seen to set forth on a botanizing expedition through the countryside.

This friendship, which was not without profit for Fabre (4/29.), was still more precious to Mill, who found, in the society of the naturalist, a certain relief from his sorrow. The substance of their conversation was far from being such as one might have imagined it. Mill was not highly sensible to the festival of nature or the poetry of the fields. He was hardly interested in botany, except from the somewhat abstract point of view of classification and the systematic arrangement of species. Always melancholy, cold, and distant, he spoke little; but Fabre felt under this apparent sensibility a rigorous integrity of character, a great capacity for devotion, and a rare goodness of heart.

So the two wandered across country, each thinking his own thoughts, and each self-contained as though they were walking on parallel but distant paths.

CHAPTER 4.

However, Fabre was not at the end of his troubles; and secret ill-feeling began to surround him. The free lectures at Saint-Martial offended the devout, angered the sectaries, and excited the intolerance of the pedants, "whose feeble eyelids blink at the daylight," and he was far from receiving, from his colleagues at the lycée, the sympathy and encouragement which were, at this moment especially, so necessary to him. Some even went so far as to denounce him publicly, and he was mentioned one day from the height of the pulpit, to the indignation of the pupils of the upper Normal College, as a man at once dangerous and subversive.

Some found it objectionable that this "irregular person, this man of solitary study," should, by his work and by the magic of his teaching, assume a position so unique and so disproportionate. Others regarded the novelty of placing the sciences at the disposal of young girls as a heresy and a scandal.

Their bickering, their cabals, their secret manoeuvres, were in the long run to triumph. Duruy had just succumbed under the incessant attacks of the clericals. In him Fabre lost a friend, a protector, and his only support. Embittered, defeated, he was now only waiting for a pretext, an incident, a mere nothing, to throw up everything.

CHAPTER 4.

One fine morning his landladies, devout and aged spinsters, made themselves the instruments of the spite of his enemies, and abruptly gave him notice to quit. he had to leave before the end of the month, for, simple and confident as usual, he had obtained neither a lease nor the least written agreement.

At this moment he was so poor that he had not even the money to meet the expenses of his removal. The times were troublous: the great war had commenced, and Paris being invested he could no longer obtain the small earnings which his textbooks were beginning to yield him, and which had for some time been increasing his modest earnings. On the other hand, having always lived far from all society, he had not at Avignon a single relation who could assist him, and he could neither obtain credit nor find any one to extricate him from his embarrassments and save him from the extremity of need with which he was threatened. He thought of Mill, and in this difficult juncture it was Mill who saved him. The philosopher was then in England; he was for the time being a member of the House of Commons, and he used to vary his life at Avignon by a few weeks' sojourn in London. His reply, however, was not long in coming: almost immediately he sent help; a sum of some 120 pounds sterling, which fell like manna into the hands of Fabre; and he did not, in exchange, demand the slightest security for this advance.

CHAPTER 4.

Then, filled with disgust, the "irregular person" shook off the yoke and retired to Orange. At first he took shelter where he could, anxious only to avoid as far as possible any contact with his fellow-men; then, having finally discovered a dwelling altogether in conformity with his tastes, he moved to the outskirts of the city, and settled at the edge of the fields, in the middle of a great meadow, in an isolated house, pleasant and commodious, connected with the road to Camaret by a superb avenue of tall and handsome plane-trees. This hermitage in some respects recalled that of Mill in the outskirts of Avignon; and thence his eyes, embracing a vast horizon, from the pediment of the ancient theatre to the hills of Sérignan, could already distinguish the promised land.

CHAPTER 5.

A GREAT TEACHER.

It was in 1871. Fabre had lived twenty years at Avignon. This date constitutes an important landmark in his career, since it marks the precise moment of his final rupture with the University.

At this time the preoccupations of material life were more pressing than ever, and it was then that he devoted himself entirely and with perseverance to the writing of those admirable works of introduction and initiation, in which he applied himself to rendering science accessible to the youngest minds, and employed all his profound knowledge to the thorough teaching of its elements and its eternal laws.

To this ungrateful task--ungrateful, but in reality pleasurable, so strongly had he the vocation, the feeling, and the genius of the teacher-- Fabre applied himself thenceforth with all his heart, and for nine years never lifted his hand.

How insipid, how forbidding were the usual classbooks, the second-rate natural histories above all, stuffed with dry statements, with raw knowledge, which brought nothing but the memory into play! How many youthful faces had grown pale above them!

What a contrast and a deliverance in these little books of Fabre's, so clear, so luminous, so simple, which for the first time spoke to the heart and the understanding; for "work which one does not understand disgusts one." (5/1.)

To initiate others into science or art, it is not enough to have understood them oneself; it is not enough even that one should be an artist or a scientist. Scientists of the highest flight are sometimes very unskilful teachers, and very indifferent hands at explaining the alphabet. It is not given to the first comer to educate the young; to understand how to identify his understanding with theirs, to measure their powers. It is a matter of instinct and good sense rather than of memory or erudition, and Fabre, who had never in his life been the pupil of any one, could better than any remember the phases through which his mind had passed, could recollect by what detours of the mind, by what secret labours of thought, by what intuitive methods he had succeeded in conquering, one by one, all the difficulties in his path, and in gradually attaining to knowledge.

It is wonderful to watch the mastery with which he conducts his demonstrations, the simplest as well as the most involved, singling out the essential, little by little evoking the sense of things, ingeniously seeking familiar examples, finding comparisons, and employing picturesque and striking images, which throw a dazzling light upon the obscurest question or the most difficult problem. How in such matters can one dispense with figurative speech, when one is reduced, as a rule, to an inability to show the things themselves, but only their images and their symbols?

Follow him, for example, in the "The Sky" (5/2.), which seems to thrill with the ardent and comprehensive genius of a Humboldt, and admire the ease with which he surmounts all the difficulties and smooths the way for the vast voyage on which he conducts you, past the infinity of the suns and the stars in their millions, scintillating in the cold air of night, to descend once more to our humble "Earth" (5/3.); first an ocean of fire, rolling its heavy waves of molten porphyry and granite, then "slowly hardening into strange floes and bergs, hotter than the red iron in the fire of the forge," rounding its back, all covered with gaping pustules, eruptive mountains and craters, and the first folds of its calcined crust, until the day when the vast mist of densest vapours, heaped up on every hand and of immeasurable depth, begins gradually to show rifts, giving rise at last to an infinite storm, a stupendous deluge, and forming the strange universal sea, "a mineral sludge, veiled by a chaos of smoke," whence at length the primitive soil emerges, "and at last the green grass."

And although "a little animal proteid, capable of pleasure and pain, surpasses in interest the whole immense creation of dead matter," he does not forget to show us the spectacle of life flowing through matter itself; and he animates even the simple elementary bodies, celebrating the marvellous activities of the air, the violence of Chlorine, the metamorphoses of Carbon, the miraculous bridals of Phosphorus, and "the splendours which accompany the birth of a drop of water." (5/4.)

A man must indeed love knowledge deeply before he can make others love it, or render it easy and attractive, revealing only the smiling highways; and Fabre, above all things the impassioned professor, was the very man to lead his disciples "between the hedges of hawthorn and sloe," whether to show them the sap, "that fruitful current, that flowing flesh, that vegetable blood," or how the plant, by a mysterious transubstantiation, makes its wood, "and the delicate bundle of swaddling-bands of its buds," or how "from a putrid ordure it extracts the flavour and the fragrance of its fruits"; or whether he seeks to evoke the murderous plants that live as parasites at the cost of others; the white Clandestinus, "which strangles the roots of the alders beside the rivers," the Cuscuta, "which knows nothing of labour," the wicked Orobanche, plump, powerful and brazen, the skin covered with ugly scales, "with sombre flowers that wear the livery of death, which leaps at the throat of the clover, stifling it, devouring it, sucking its blood." (5/5.)

Botany, by this genial treatment, becomes a most interesting study, and I know of no more captivating reading than "The Plant" and "The Story of the Log," the jewels of this incomparable series.

Employ Fabre's method if you wish to learn by yourself, or to evoke in your children a love of science, and, according to the phrase of the gentle Jean-Jacques, to help them "to buy at the best possible of prices." Give them as sole guides these exquisite manuals, which touch upon everything, initiating them into everything, and bringing within the reach of all, for their instruction or amusement, the heavens and the earth, the planets and their moons, the mechanism of the great natural forces and the laws which govern them, life and its materials, agriculture and its applications. For more than a quarter of a century these catechisms of science, models of lucidity and good sense, effected the education of generations of Frenchmen. Abridgments of all knowledge, veritable codes of rural wisdom, these perfect breviaries have never been surpassed.

It was after reading these little books, it is said, that Duruy conceived the idea of confiding to this admirable teacher the education of the Imperial heir; and it is very probable that this was, in reality, the secret motive which would explain why he had so expressly summoned Fabre to Paris. What an ideal tutor he had thought of, and how proud might others have been of such a choice! But the man was too zealous of his independence, too difficult to tame, to bear with the environment of a court, and God knows whether he was made for such refulgence! We need not be surprised that Fabre never heard of it; it must have sufficed the minister to speak with him for a few minutes to realize that the most tempting offers and all the powers of seduction would never overcome his insurmountable dislike of life in a capital, nor prevail against his inborn, passionate, exclusive love of the open.

For these volumes Fabre was at first rather wretchedly paid; at all events, until public education had definitely received a fresh impulse; and for a long time his life at Orange was literally a hand-to-mouth existence.

As soon as he was able to realize a few advances, he had nothing so much at heart as the repayment of Mill, and he hastened to call on the philosopher; all the more filled with gratitude for his generosity in that the loan, although of the comparatively large amount of three thousand francs, was made without security, practically from hand to hand, with no other warranty than his probity.

For this reason this episode was always engraven on his memory. Thirty years later he would relate the affair even to the most insignificant details. How many times has he not reminded me of the transaction, insisting that I should make a note of it, so anxious was he that this incident in his career should not be lost in oblivion! How often has he not recalled the infinite delicacy of Mill, and his excessive scrupulousness, which went so far that he wished to give a written acknowledgment of the repayment of the debt, of which there was no record whatever save in the conscience of the debtor!

Scarcely two years later Mill died suddenly at Avignon. Grief finally killed him; for this unexpected death seemed to have been only the ultimate climax of the secret malady which had so long been undermining him.

It was in the outskirts of Orange that Fabre for the last time met him and accompanied him upon a botanizing expedition. He was struck by his weakness and his rapid decline. Mill could hardly drag himself along, and when he stooped to gather a specimen he had the greatest difficulty in rising. They were never to meet again.

CHAPTER 5.

A few days later--on the 8th May, 1873--Fabre was invited to lunch with the philosopher. Before going to the little house by the cemetery he halted, as was his custom, at the Libraire Saint-Just. It was there that he learned, with amazement, of the tragic and sudden event which set a so unexpected term to a friendship which was doubtless a little remote, but which was, on both sides, a singularly lofty and beautiful attachment.

His class-books were now bringing in scarcely anything; their preparation, moreover, involved an excessive expenditure of time, and gave him a great deal of trouble; it is impossible to imagine what scrupulous care, what zeal and self-respect Fabre brought to the execution of the programme which he had to fulfil.

To begin with, he considered that he could not enjoy a more splendid opportunity to give children a taste for science and to stimulate their curiosity than by finding a means to interest them, from their earliest infancy, in their simple playthings, even the crudest and most inexpensive; so true is it that "in the smallest mechanical device or engine, even in its simplest form, as conceived by the industry of a child, there is often the germ of important truths, and, better than books, the school of the playroom, if gently disciplined, will open for the child the windows of the universe."

"The humble teetotum, made of a crust of rye-bread transfixed by a twig, silently spinning on the cover of a school-book, will give a correct enough image of the earth, which retains unmoved its original impulse, and travels along a great circle, at the same time turning on itself. Gummed on its disc, scraps of paper properly coloured will tell us of white light, decomposable into various coloured rays...

"There will be the pop-gun, with its ramrod and its two plugs of tow, the hinder one expelling the foremost by the elasticity of the compressed air. Thus we get a glimpse of the ballistics of gunpowder, and the pressure of steam in engines..."

The little hydraulic fountain made of an apricot stone, patiently hollowed and pierced with a hole at either side, into which two straws are fitted, one dipping into a cup of water and the other duly capped, "expelling a slender thread of water in which the sunlight flickers," will introduce us to the true syphon of physics.

"What amusing and useful lessons" a well-balanced scheme of education might extract from this "academy of childish ingenuity"! (5/6.)

CHAPTER 5.

At this time he was undertaking the education of his own children. His chemistry lessons especially had a great success. (5/7.) With apparatus of his own devising and of the simplest kind, he could perform a host of elementary experiments, the apparatus as a rule consisting of the most ordinary materials, such as a common flask or bottle, an old mustard-pot, a tumbler, a goose-quill or a pipe-stem.

A series of astonishing phenomena amazed their wondering eyes. He made them see, touch, taste, handle, and smell, and always "the hand assisted the word," always "the example accompanied the precept," for no one more fully valued the profound maxim, so neglected and misunderstood, that "to see is to know."

He exerted himself to arouse their curiosity, to provoke their questions, to discover their mistakes, to set their ideas in order; he accustomed them to rectify their errors themselves, and from all this he obtained excellent material for his books.

For those more especially intended for the education of girls he took counsel with his daughter Antonia, inviting her collaboration, begging her to suggest every aspect of the matter that occurred to her; for instance, in respect of the chemistry of the household, "where exact science should shed its light upon a host of facts relating to domestic economy" (5/8.), from the washing of clothes to the making of a stew.

Even now, to his despair, although freed from the cares of school life, he was always almost wholly without leisure to devote himself to his chosen subjects.

It was at this period above all that he felt so "lonely, abandoned, struggling against misfortune; and before one can philosophize one has to live." (5/9.)

And his incessant labour was aggravated by a bitter disappointment. In the year of Mill's death Fabre was dismissed from his post as conservator of the Requien Museum, which he had held in spite of his departure from Avignon, going thither regularly twice a week to acquit himself of his duties. The municipality, working in the dark, suddenly dismissed him without explanation. To Fabre this dismissal was infinitely bitter; "a sweeper-boy would have been treated with as much ceremony." (5/10.) What afflicted him most was not the undeserved slight of the dismissal, but his unspeakable regret at quitting those beloved vegetable collections, "amassed with such love" by Requien, who was his friend and master, and by Mill and himself; and the thought that he would henceforth perhaps be unable to save these precious but perishable things from oblivion, or terminate the botanical geography of Vaucluse, on which he had been thirty years at work!

For this reason, when there was some talk of establishing an agronomic station at Avignon, and of appointing him director, he was at first warmly in favour of the idea. (5/11.) Already he foresaw a host of fascinating experiments, of the highest practical value, conducted in the peace and leisure and security of a fixed appointment. It is indeed probable that in so vast a field he would have demonstrated many valuable truths, fruitful in practical results; he was certainly meant for such a task, and he would have performed it with genuine personal satisfaction. He had already exerted his ingenuity by trying to develop, among the children of the countryside, a taste for agriculture, which he rightly considered the logical complement of the primary school, and which is based upon all the sciences which he himself had studied, probed, taught, and popularized.

It will be remembered how patiently he devoted himself for twelve years to the study of madder, multiplying his researches, and applying himself not only to extracting the colouring principle, but also to indicating means whereby adulteration and fraud might be detected.

He had published memoirs of great importance dealing with entomology in its relations to agriculture. Impressed with the importance of this little world, he suggested valuable remedies, means of preservation; which were all the more logical in that the destruction of insects, if it is to be efficacious, must be based not upon a gross empiricism, but on a previous study of their social life and their habits.

With what patience he observed the terribly destructive weevils, and those formidable moths with downy wings, which fly without sound of a night, and whose depredations have often been valued at millions of francs! How meticulously he has recorded the conditions which favour or check the development of those parasitic fungi whose mortal blemishes are seen on buds and flowers, on the green shoots and clusters that promise a prosperous vintage!

But then he became anxious. Was it all worth the sacrifice of his liberty? "Would he not suffer a thousand annoyances from pretentious nobodies?" for as things were, all ideas of again "enregimenting" himself "filled him with horror." (5/12.)

Slowly, however, the first instalment of the work which he had spent nearly twenty-five years in planning, creating, and polishing, began to take shape. At the end of the year 1878 he was able to assemble a sufficient number of

CHAPTER 5.

studies to form material for what was to be the first volume of his "Souvenirs entomologiques." (A selection of which forms "Social Life in the Insect World" (T. Fisher Unwin, 1912).)

Let us stop for a moment to consider this first book, whose publication constitutes a truly historical date, not only in the career of Fabre, but in the annals of universal science. It was at once the foundation and the keystone of the marvellous edifice which we shall watch unfolding and increasing, but to which the future was in reality to add nothing essential. The cardinal ideas as to instinct and evolution, the necessity of experimenting in the psychology of animals, and the harmonic laws of the conservation of the individual, are here already expounded in their final and definite form. This fruitful and decisive year brought Fabre a great grief. He lost his son Jules, that one of all his children whom he seems most ardently to have loved.

He was a youth of great promise, "all fire, all flame"; of a serious nature; an exquisite being, of a precocious intelligence, whose rare aptitudes both for science and literature were truly extraordinary. Such too was the subtlety of his senses that by handling no matter what plant, with his eyes closed, he could recognize and define it merely by the sense of touch. This delightful companion of his father's studies had scarcely passed his fifteenth year when death removed him. A terrible void was left in his heart, which was never filled. Thirty years later the least allusion to this child, however tactful, which recalled this dear memory to his mind, would still wring his heart, and his whole body would be shaken by his sobs. As always, work was his refuge and consolation; but this terrible blow shattered his health, until then so robust. In the midst of this disastrous winter he fell seriously ill. He was stricken with pneumonia, which all but carried him off, and every one gave him up for lost. However, he recovered, and issued from his convalescence as though regenerated, and with strength renewed he attacked the next stage of his labours.

But what are the most fruitful resolutions, and what poor playthings are we in the hands of the unexpected! A vulgar incident of every-day life had sufficed to make Fabre decide to break openly with the University, and to leave Avignon. The secret motive of his departure from Orange was scarcely more solid. His new landlord concluded one day, either from cupidity or stupidity, to lop most ferociously the two magnificent rows of plane-trees which formed a shady avenue before his house, in which the birds piped and warbled in the spring, and the cicadae chorused in the summer. Fabre could not endure this massacre, this barbarous mutilation, this crime against nature. Hungry for peace and quiet, the enjoyment of a dwelling-place could no longer content him; at all costs he must own his own home.

So, having won the modest ransom of his deliverance, he waited no longer, but quitted the cities for ever; retiring to Sérignan, to the peaceful obscurity of a tiny hamlet, and this quiet corner of the earth had henceforth all his heart and soul in keeping.

CHAPTER 6.

THE HERMITAGE.

Goethe has somewhere written: Whosoever would understand the poet and his work should visit the poet's country.

Let us, then, the latest of many, make the pilgrimage which all those who are fascinated by the enigma of nature will accomplish later, with the same piety that has led so many and so fervent admirers to the dwelling of Mistral at Maillane.

Starting from Orange and crossing the Aygues, a torrent whose muddy waters are lost in the Rhône, but whose bed is dried by the July and August suns, leaving only a desert of pebbles, where the Mason-bee builds her pretty turrets of rock-work, we come presently to the Sérignaise country; an arid, stony tract, planted with vines and olives, coloured a rusty red, or touched here and there with almost a hue of blood; and here and there a grove of cypress makes a sombre blot. To the north runs a long black line of hills, covered with box and ilex and the giant heather of the south. Far in the distance, to the east, the immense plain is closed in by the wall of Saint-Amant and the ridge of the Dentelle, behind which the lofty Ventoux rears its rocky, cloven bosom abruptly to the clouds. At the end of a few miles of dusty road, swept by the powerful breath of the mistral, we suddenly reach a little village. It is a curious little community, with its central street adorned by a double row of plane-trees, its leaping fountains, and its almost Italian air. The houses are lime-washed, with flat roofs; and sometimes, at the side of some small or decrepit dwelling, we see the unexpected curves of a loggia. At a distance the facade of the church has the harmonious lines of a little antique temple; close at hand is the graceful campanile, an old octagonal tower surmounted by a narrow mitre wrought in hammered iron, in the midst of which are seen the black profiles of the bells.

I shall never forget my first visit. It was in the month of August; and the whole countryside was ringing with the song of the cicadae. I had applied to a job-master of Orange, counting on him to take me thither; but he had never driven any one to Sérignan, had hardly heard of Fabre, and did not know where his house was. At length, however, we contrived to find it. At the entrance of the little market-town, in a solitary corner, in the centre of an enclosure of lofty walls, which were taller than the crests of the pines and cypresses, his dwelling was hidden away. No sound proceeded from it; but for the baying of the faithful Tom I do not think I should have dared to knock on the great door, which turned slowly on its hinges. A pink

CHAPTER 6.

house with green shutters, half-hidden amid the sombre foliage, appears at the end of an alley of lilacs, "which sway in the spring under the weight of their balmy thyrsi." Before the house are the shady plane-trees, where during the burning hours of August the cicada of the flowering ash, the deafening cacan, concealed beneath the leaves, fills the hot atmosphere with its eager cries, the only sound that disturbs the profound silence of this solitude.

Before us, beyond a little wall of a height to lean upon, on an isolated lawn, beneath the shade of great trees with interwoven boughs, a circular basin displays its still surface, across which the skating Hydrometra traces its wide circles. Then, suddenly, we see an opening into the most extraordinary and unexpected of gardens; a wild park, full of strenuous vegetation, which hides the pebbly soil in all directions; a chaos of plants and bushes, created throughout especially to attract the insects of the neighbourhood.

Thickets of wild laurel and dense clumps of lavender encroach upon the paths, alternating with great bushes of coronilla, which bar the flight of the butterfly with their yellow-winged flowers, and whose searching fragrance embalms all the air about them.

CHAPTER 6.

It is as though the neighbouring mountain had one day departed, leaving here its thistles, its dogberry-trees, its brooms, its rushes, its juniper- bushes, its laburnums, and its spurges. There too grows the "strawberry tree," whose red fruits wear so familiar an appearance; and tall pines, the giants of this "pigmy forest." There the Japanese privet ripens its black berries, mingled with the Paulownia and the Cratoegus with their tender green foliage. Coltsfoot mingles with violets; clumps of sage and thyme mix their fragrance with the scent of rosemary and a host of balsamic plants. Amid the cacti, their fleshy leaves bristling with prickles, the periwinkle opens its scattered blossoms, while in a corner the serpent arum raises its cornucopia, in which those insects that love putrescence fall engulfed, deceived by the horrible savour of its exhalations.

It is in the spring above all that one should see this torrent of verdure, when the whole enclosure awakens in its festival attire, decked with all the flowers of May, and the warm air, full of the hum of insects, is perfumed with a thousand intoxicating scents. It is in the spring that one should see the "Harmas," the open-air observatory, "the laboratory of living entomology" (6/1.); a name and a spot which Fabre has made famous throughout the world.

I enter the dining-room, whose wide, half-closed shutters allow only a half-light to enter between the printed curtains.

CHAPTER 6.

Rush-bottomed chairs, a great table, about which seven persons daily take their places, a few poor pieces of furniture, and a simple bookcase; such are all the contents. On the mantel, a clock in black marble, a precious souvenir, the only present which Fabre received at the time of his exodus from Avignon; it was given by his old pupils, the young girls who used to attend the free lectures at Saint-Martial's.

There, every afternoon, half lying on a little sofa, the naturalist has the habit of taking a short siesta. This light repose, even without sleep, was of old enough to restore his energies, exhausted by hours of labour. Thenceforth he was once more alert, and ready for the remainder of the day.

But already he is on his feet, bareheaded, in his waistcoat, his silk necktie carelessly fastened under the soft turned-down collar of his half- open shirt, his gesture, in the shadowy chamber, full of welcome.

François Sicard, in his faultless medal and his admirable bust, has succeeded with rare felicity in reproducing for posterity this rugged, shaven face, full of laborious years; a peasant face, stamped with originality, under the wide felt hat of Provence; touched with geniality and benevolence, yet reflecting a world of energy. Sicard has fixed for ever this strange mask; the thin cheeks, ploughed into deep furrows, the strained nose, the pendent wrinkles of the throat, the thin, shrivelled lips, with an indescribable fold of bitterness at the corners of the mouth. The hair, tossed back, falls in fine curls over the ears, revealing a high, rounded forehead, obstinate and full of thought. But what chisel, what graver could reproduce the surprising shrewdness of that gaze, eclipsed from time to time by a convulsive tremor of the eyelids! What Holbein, what Chardin could render the almost extraordinary brilliance of those black eyes, those dilated pupils: the eyes of a prophet, a seer; singularly wide and deeply set, as though gazing always upon the mystery of things, as though made expressly to scrutinize Nature and decipher her enigmas? Above the orbits, two short, bristling eyebrows seem set there to guide the vision; one, by dint of knitting itself above the magnifying-glass, has retained an indelible fold of continual attention; the other, on the contrary, always updrawn, has the look of defying the interlocutor, of foreseeing his objections, of waiting with an ever-ready return-thrust. Such is this striking physiognomy, which one who has seen it cannot forget.

There, in this "hermit's retreat," as he himself has defined it, the sage is voluntarily sequestered; a true saint of science, an ascetic living only on fruits, vegetables, and a little wine; so in love with retirement that even in the village he was for a long time almost unknown, so careful was he to go round instead of through it on his way to the neighbouring mountain, where he would often spend whole days alone with wild nature.

It is in this silent Thebaïd, so far from the atmosphere of cities, the vain agitations and storms of the world, that his life has been passed, in unchanging uniformity; and here he has been able to pursue, with resolute labour and incredible patience, that prodigious series of marvellous observations which for nearly fifty years he has never ceased to accumulate.

Let us indeed remember how much time has been required and what effort has been expended to complete the long and patient inquiries which he had hitherto accomplished; obliged, as he was, to allow himself to be interrupted at any moment, and to postpone his observations often at the most interesting moment, in order to undertake some enervating labour, or the disagreeable and mechanical duties of his profession. Remember that his first labours already dated from twenty-five years earlier, and at the moment when we observe him in his solitude at Sérignan he had only just painfully gathered together the material for his first book. What a contrast to the thirty fruitful years that were to follow! Now nearly ten volumes, no less overflowing with the richest material, were to succeed one another at almost regular intervals--about one in every three years.

To be sure, he would have gathered his harvest in no matter what corner of the world, provided he had found within his reach, in whatever sphere of life he had been placed, any subject of inquiry whatever; such was Rousseau, botanizing over the bunch of chickweed provided for his canary; such was Bernardin Saint-Pierre, discovering a world in a strawberry-plant which had sprouted by chance at the corner of his window. (6/2.) But the field in which he had hitherto been able to glean was indeed barren. That he was able, later on, to narrate the wonderful history of the Pelopaeus, whose habits he had observed at Avignon, was due to the fact that this curious insect had come to lodge with him, having chosen Fabre's chamber for its dwelling. None the less he threw himself eagerly upon all such scraps of information as happened to come under his notice; witness the observations which he embodied in a memoir touching the phosphorescence of certain earth-worms which, abounding in a little courtyard near his dwelling, were so rare elsewhere that he was never again able to find them. (6/3.) It was therefore fortunate, if not for himself, at least for his genius, that he did not become, as he had wished, a professor in a faculty; there, to be sure, he would have found a theatre worthy of his efforts, in which he might even have demonstrated, in all its magnificence, his incomparable gift of teaching; but it is probable too that he would have been stranded in shoal waters; that in the official atmosphere of a city his still more marvellous gifts of observation would scarcely have found employment.

It was only by belonging fully to himself that he could fruitfully exercise his talents. Necessary to every scholar, to every inquirer, to an open-air observer like Fabre liberty and leisure were more than usually essential; failing these he might never have accomplished his mission. How many lives are wasted, how many minds expended in sheer loss, in default of this sufficiency of leisure! How many scholars tied to the soil, how many physicians absorbed by an exigent practice, who perhaps had somewhat to say, have succeeded only in devising plans, for ever postponing their realization to some miraculous tomorrow, which always recedes!

But we must not fall into illusions. How many might be tempted to imitate him, hoping to see some unknown talent awaken or expand within them, only to find themselves incapable of producing anything, and to consume themselves in an insurmountable and barren ennui! One must be rich in one's own nature, rich in will and in ability, to live apart and seek new paths in solitude, and it is not without reason that the majority prefer the turmoil of cities and the murmur of men to the silence of the country.

The atmosphere of a great capital, for instance, is singularly conducive to work. Living constantly within the circle of light shed by the masters, within reach of the laboratories and the great libraries, we are less likely to go astray; we are stimulated by the contact of others; we profit by their advice and experience; and it is easy to borrow ideas if we lack them. Then there is the stimulant of self-respect, the sense of rivalry, the eager desire to advance, to distinguish oneself, to shine, to attract attention, to become in one's turn an arbiter, an object of wonder and envy, without which stimulus many would merely have existed, and would never have become what they are.

On the other hand, a man needs an intrinsic radio-activity, and a real talent; and the aid, moreover, of exceptional circumstances, if fame is to consent to come to him and take him by the hand in the depths of some unknown Maillane, some obscure Sérignan; even, as in the case of Fabre, at the end only of a long life.

But he, by a kind of fatality inherent in his nature, loved "to circumscribe himself," according to the happy expression of Rousseau; and he profited, rather than otherwise, by living entirely to himself; for he had long been, indeed he

CHAPTER 6.

always was, the man who, at twenty-five, writing to his brother, had said, in speaking of his native countryside:

"For a impassioned botanist, it is a delightful country, in which I could pass a month, two months, three months, a year even, alone, quite alone, with no other companion than the crows and the jays which gossip among the oak-trees; without being weary for a moment; there would be so many beautiful fungi, orange, rosy, and white, among the mosses, and so many flowers in the fields." (6/4.)

His work having brought him at last just enough to enable him to give himself the pleasure of becoming, in his turn, a proprietor, he had acquired, for a modest sum, this dilapidated dwelling and this deserted spot of ground; barren land, given over to couch-grass, thistles, and brambles; a sort of "accursed spot, to which no one would have confided even a pinch of turnip-seed." A piece of water in front of the house attracted all the frogs in the neighbourhood; the screech-owl mewed from the tops of the plane-trees, and numerous birds, no longer disturbed by the presence of man, had domiciled themselves in the lilacs and the cypresses. A host of insects had seized upon the dwelling, which had long been deserted.

He restored the house, and to some extent reduced confusion to order. In the uncultivated and pebbly plain where the plough had been long a stranger he established plants of a thousand varieties, and, the better to hide himself, he had walls built to shut himself in.

Why was he drawn by preference to this village of Sérignan?--for he did not go thither without making some inquiries as to the possibility of obtaining shelter elsewhere, and the Carpentras cemetery had tempted him also; but what had particularly seduced and drawn him thither was the nearness of the mountain with its Mediterranean flora, so rich that it recalled the Corsican maquis; full of beautiful fungi and varied insects, where, under the flat stones exposed to the burning sun, the centipede burrowed and the scorpion slept; where a special fauna abounded--of curious dung-beetles, scarabaei, the Copris, the Minotaur, etc.--which only a little farther north grow rapidly scarcer and then altogether disappear.

He had thus at last arrived in port; he had found his "Eden."

He had realized, "after forty years of desperate struggles," the dearest, the most ardent, the longest cherished of all his desires. He could observe at leisure "every day, every hour," his beloved insects; "under the blue sky, to the music of the cigales." He had only to open his eyes and to see; to lend an ear and hear; to enjoy the great blessing of leisure to his heart's content.

Doffing the professor's frock-coat for the peasant's blouse, planting a root of sweet basil in his "topper," and finally kicking it to pieces, he snapped his fingers at his past life.

Liberated at last, far from all that could irritate or disturb him or make him feel dependent, satisfied with his modest earnings, reassured by the ever-increasing popularity of his little books, he had obtained entire possession of his own body and mind, and could give himself without reserve to his favourite subjects.

So, with Nature and her inexhaustible book before him, he truly commenced a new life.

But would this life have been possible without the support and comfort of those intimate feelings which are at the root of human nature? Man is seldom the master of these feelings, and they, with reason or despite reason, force themselves on his notice as the question of questions.

This delicate problem Fabre had to resolve after suffering a fresh grief. Hardly had he commenced to enjoy the benefits of this profound peace, when he lost his wife. At this moment his children were already grown up; some were married and some ready to leave him; and he could not hope much longer to keep his old father, the ex-café-keeper of Pierrelatte, who had come to rejoin him; and who might be seen, even in his extreme old age, going forth in all weathers and dragging his aged limbs along all the roads of Sérignan. (6/5.) The son, moreover, had inherited from his father his profound inaptitude for the practical business of life, and was equally incapable of managing his interests and the economics of the house. This is why, after two years of widowerhood, having already passed his sixtieth year, although still physically quite youthful, he remarried. Careless of opinion, obeying only the dictates of his own heart and mind, and following also the intuitions of unerring instinct, which was superior to the understanding of those who thought it their duty to oppose him, he married, as Boaz married Ruth, a young woman, industrious, full of freshness and life, already completely devoted to his service, and admirably fitted to satisfy that craving for order, peace, quiet, and moral tranquillity, which to him were above all things indispensable.

His new companion, moreover, was in all things faithful to her mission, and it was thanks to the benefits of this union, as the future was to show, that Fabre was in a position to pursue his long-delayed inquiries.

CHAPTER 6.

Three children, a son and two daughters, were born in swift succession, and reconstituted "the family," which was very soon increased by the youngest of his daughters by his first wife, who had not married; this was that Aglaë, who so often helped her father with her childlike attentions, and, "her cheek blooming with animation," collaborated in some of his most famous observations (6/6.); an unobtrusive figure, a soul full of devotion and resignation, heroic and tender. Having in vain ventured into the world, she had returned to the beloved roof at Sérignan, unable to part from the father she so admired and adored.

Later, when the shadow of age grew denser and heavier, the young wife and the younger children of the famous poet-entomologist took part in his labours also; they gave him their material assistance, their hands, their eyes, their hearing, their feet; he in the midst of them was the conceiving, reasoning, interpreting, and directing brain.

>From this time forward the biography of Fabre becomes simplified, and remains a statement of his inner life. For thirty years he never emerged from his horizon of mountains and his garden of shingle; he lived wholly absorbed in domestic affections and the tasks of a naturalist. None the less, he still exercised his vocation as teacher, for neither pure science nor poetry was sufficient to nourish his mind, and he was still Professor Fabre,

CHAPTER 6.

untiringly pursuing his programme of education, although no longer applying himself thereto exclusively.

This long active period was also the most silent period of his life, although not an hour, not a minute of his many days was left unoccupied.

In the first few months at his new home he resumed his hymn to labour.

"You will learn in your turn," he writes to his son Émile, "you will learn, I hope, that we are never so happy as when work does not leave us a moment's repose. To act is to live." (6/7.)

The better to belong to himself, he eluded all invitations, even those from his nearest or most intimate friends; he hated to go away even for a few hours, preferring to enjoy in his own house their presence amidst his habitual and delightful surroundings. Everything in this still unexplored country was new to him. What would he do elsewhere, even in his beloved Carpentras, whither his faithful friend and pupil Devillario, who had formerly followed him in his walks around Avignon, would endeavour from time to time to draw him? Devillario was a magistrate, a collector and palaeontologist; his simple tastes, his wide culture, and his passion for natural history would surely have decided

Fabre to accept his invitations, but that he forbade himself the pleasure. "I am afraid the hospitable cutlet that awaits me at your table will have time to grow cold; I am up to the neck in my work (6/8.)...But you, when you can, escape from your courts, and we will philosophize at random, as is our custom when we can manage to pass a few hours together. As for me, it is very doubtful whether the temptation will seize me to come to Carpentras. A hermit of the Thebaïd was no more diligent in his cell than I in my village home." (6/9.)

CHAPTER 7.

THE INTERPRETATION OF NATURE.

Was there not indeed a sufficiency of captivating matters all about him, and beneath his very feet?

In his deep, sunny garden a thousand insects fly, creep, crawl, and hum, and each relates its history to him. A golden gardener-beetle trots along the path. Rose-beetles pass, in snoring flight, on every hand, the gold and emerald of their elytra gleaming; now and again one of them alights for a moment on the flowering head of a thistle; he seizes it carefully with the tips of his nervous, pointed fingers, seems to caress it, speaks to it, and then suddenly restores it to freedom.

Wasps are pillaging the centauries. On the blossoms of the camomile the larvae of the Meloë are waiting for the Anthophorae to carry them off to their cells, while around them roam the Cicindelae, their green bodies "spotted with points of amaranth." At the bottom of the walls "the chilly Psyche creeps slowly along under her cloak of tiny twigs." In the dead bough of a lilac-tree the dark-hued Xylocopa, the wood-boring bee, is busy tunnelling her gallery. In the shade of the rushes the Praying Mantis, rustling the floating robe of her long tender green wings, "gazes alertly, on the watch, her arms folded on her breast, her appearance that of one praying," and paralyses the great grey locust, nailed to its place by fear.

Nothing here is insignificant; what the world would smile at or deride will provide the sage with food for thought and reflection. "Nothing is trivial in the majestic problem of nature; our laboratory acquaria are of less value than the imprint which the shoe of a mule has left in the clay, when the rain has filled the primitive basin, and life has peopled it with marvels"; and the least fact offered us by chance on the most thoroughly beaten track may possibly open prospects as vast as all the starry sky.

Tell yourself that everything in nature is a symbol of something like a specimen of an abstruse cryptogram, all the characters of which conceal some meaning. But when we have succeeded in deciphering these living texts, and have grasped the allusion; when, beside the symbol, we have succeeded in finding the commentary, then the most desolate corner of the earth appears to the solitary seeker as a gallery full of the masterpieces of an unsuspected art. Fabre puts into our hands the golden key which opens the doors of this marvellous museum.

Let us consider the terebinth louse; it is just a little yellow mite; but is it nothing else? Its genealogical history teaches us "by what amazing essays of passion and variety the universal law which rules the transmission of life is evolved. Here is neither father nor eggs; all these mites are mothers; and the young are born living, just like their mothers." To this end "almost the whole of the maternal substance is disintegrated and renewed and conglobated to form the ovarium...the whole creature has become an egg, which has, for its shell, the dry skin of the tiny creature, and the microscope will show a whole world in formation...a nebulosity as of white of egg, in which fresh centres of life are forming, as the suns are condensed in the nebulae of the heavens." (7/1.)

What is this fleck of foam, like a drop of saliva, which we see in springtime on the weeds of the meadows; among others on the spurge, when its stems begin to shoot, and its sombre flowers open in the sunlight? "It is the work of an insect. It is the shelter in which the Cicadellina deposits her eggs. What a miraculous chemist! Her stiletto excels the finest craft of the botanical anatomist" by its sovereign art of separating the acrid poison which flows with the sap in the veins of the most venomous plants, and extracting therefrom only an inoffensive fluid. (7/2.)

At every step the insects set us problems equally varied. The other creatures are nearer to us; they resemble us in many respects. But insects, almost the first-born of creation, form a world apart, and contain, in their tiny bodies, as Réaumur has admirably said, "more parts than the most gigantic animals." They have senses and faculties of their own, which enable them to accomplish actions, which are doubtless very simply related in reality, but which seem, to our minds, as extraordinary as the habits of the inhabitants of Mars might, if by chance they were to descend in our midst. We do not know how they hear, nor how they see through their compound eyes, and our ignorance concerning the majority of their senses still further increases the difficulty, which so often arrests us, of interpreting their actions.

The tubercled Cerceris "finds by the hundred" and almost immediately a species of weevil, the Cleona ophthalmica, on which it feeds its larvae, and which the human eye, though it searches for hours, can scarcely find anywhere. The eyes of the Cerceris are like magnifying glasses, veritable microscopes, which immediately distinguish, in the vast field of nature, an object that human vision is powerless to discover. (7/3.)

How does the Ammophila, hovering over the turf and investigating it far and wide, in its search for a grey grub, contrive to discern the precise point in the depth of the subsoil where the larva is slumbering in immobility? "Neither touch nor sight can come into play, for the grub is sealed up in its burrow at a depth of several inches; nor the scent, since it is absolutely inodorous; nor the hearing, since its immobility is absolute during the daytime." (7/4.)

The Processional caterpillar of the pine-trees, "endowed with an exquisite hygrometric sensibility," is a barometer more infallible than that of the physicists. "It foresees the tempests preparing afar, at enormous distances, almost in the other hemisphere," and announces them several days before the least sign of them appears on the horizon. (7/5.)

A wild bee, the Chalicodoma, and a wasp, the Cerceris, carried in the dark far from their familiar pastures, to a distance of several miles, and released in spots which they have never seen, cross vast and unknown spaces with absolute certainty, and regain their nests; even after long absence, and in spite of contrary winds and the most unexpected obstacles. It is not memory that guides them, but a special faculty whose astonishing results we must admit without attempting to explain them, so far removed are they from our own psychology. (7/6.) But here is another example:

The Greater Peacock moths cross hills and valleys in the darkness, with a heavy flight of wings spotted with inexplicable hieroglyphics. They hasten from the remotest depths of the horizon to find their "sleeping beauties," drawn thereto by unknown odours, inappreciable by our senses, yet so penetrating that the branch of almond on which the female has perched, and which she has impregnated with her effluvium, exerts the same extraordinary attraction. (7/7.)

Considering these creatures, we end by discovering more things than are contained in all the philosophies...if we know how to look for them.

Among so many unimaginable phenomena, which bewilder us, "because there is nothing analogous in us," we succeed in perceiving, here and there, a few glimpses of day, which suddenly throw a singular light upon this black labyrinth, in which the least secret we can surprise "enters perhaps more directly into the profound enigma of our ends and our origins than the secret of the most urgent and most closely studied of our passions." (7/8.)

Fabre explains by hypnosis one of those curious facts which have hitherto been so poorly interpreted. When surprised by abnormal conditions, we see insects suddenly fall over, drop to the ground, and lie as though struck by lightning, gathering their limbs under their bodies. A shock, an unexpected odour, a loud noise, plunges them instantly into a sort of lethargy, more or less prolonged. The insect "feigns death," not because it simulates death, but in reality because this MAGNETIC condition resembles that of death. (7/9.) Now the Odynerus, the Anthidium, the Eucera, the Ammophila, and all the hymenoptera which Fabre has observed sleeping at the fall of night, "suspended in space solely by the strength of their mandibles, their bodies tense, their limbs retracted, without exhaustion or collapse"; and the larva of the Empusa, "which for some ten months hangs to a twig by its limbs, head downwards": do not these present a surprising analogy with those hypnotized persons who possess the faculty of remaining fixed in the most painful poses, and of supporting the most unusual attitudes, for an extremely long time; for instance, with one arm extended, or one foot raised from the ground, without appearing to experience the least fatigue, and with a persevering and unfaltering energy? (7/10.)

That the ex-schoolmaster was able to penetrate so far into this new world, and that he has been able to interest us in so many fascinating problems, was due to the fact that he had also "taken a wide bird's-eye view through all the windows of creation." His universal capabilities, his immense culture and almost encyclopaedic science have enabled him to utilize, thanks to his studies, all the knowledge allied to his subject. He is not one of those who understand only their speciality and who, knowing nothing outside their own province and their particular labours, refuse to grasp at anything beyond the narrow limits within which they stand installed.

All plants are to him so familiar that the flowers, for him, assume the airs of living persons. But without a profound knowledge of botany, who would hope to grasp the profound, perpetual, and intimate relations of the plant and the insect?

He has turned over strata and interrogated the schistous deposits, whose archives preserve the forms of vanished organizations, but "keep silence as to the origin of the instincts." Bending over his reagents, he has sought to discover, according to the phrase of a philosopher, those secret retreats in which Nature is seated before her furnaces, in the depths of her laboratory; following up the metamorphoses of matter even to the wings of the Scarabaei, and observing how life, returning to her crucible the debris and ashes of the organism, combines the elements anew, and from the elements of the urine can derive, for example, by a simple displacement of molecules, "all this dazzling magic of colours of innumerable shades: the amethystine violet of Geotrupes, the emerald of the rose-beetle, the gilded green of the Cantharides, the metallic lustre of the gardener-beetles, and all the pomp of the Buprestes and the dung-beetles." (7/11.)

His books are steeped in all the ideas of modern physics. The highest mathematical knowledge has been referred to with profit in his marvellous description of the hunting-net of the Epeïra. Whose "terribly scientific" combinations realize "the spiral logarithm of the geometers, so curious in its properties" (7/12.); a splendid observation, in which Fabre makes us admire, in the humble web of a spider, a masterpiece as astonishing and incomprehensible as and even more sublime than the honeycomb.

This explains why Fabre has always energetically denied that he is properly speaking an entomologist; and indeed the term appears often wrongly to describe him. He loves, on the contrary, to call himself a naturalist; that is, a biologist; biology being, by definition, the study of living creatures considered as a whole and from every point of view. And as nothing in life is isolated, as all things hold together, and as each part, in all its relations, presents itself to the gaze of the observer under innumerable aspects, one cannot be a true naturalist without being at the same time a philosopher.

But it is not enough to know and to observe.

To be admitted to the spectacle of these tiny creatures, to become familiar with their habits, to grasp the mysterious threads which connect them one with another and with the vast universe: for this the cold and deliberate vision of the specialist would often be insufficient. There is an art of observation, and the gift of observation is a true function of that constantly alert intelligence, continually dominated by the need of delving untiringly down to the ultimate truth accessible, "allowing ourselves to pass over nothing without seeking its reason, and habitually following up every response with another question, until we come to the granite wall of the Unknowable." Above all we need an ardent and interested sympathy, for "we penetrate farther into the secret of things by the heart than by the reason," as Toussenel has said; and "it is only by intuition that we can know what life truly is," adds Bergson profoundly. (7/13.) Now Fabre loves these little peoples and knows how to make us love them. How tenderly he speaks of them; with what solicitude he observes them; with what love he follows the progress of their nurslings; the young grubs wriggling in his test-tubes, with doddering heads, are happy; and he himself is happy to see them "well-fed and shining with health." He pities the bee stabbed by the Philanthus "in the holy joys of labour." He sympathizes with the sufferings of these little creatures and their hard labours. If, in his search for ideas, he has to overturn their dwellings, "he repents of subjecting maternal love to such tribulations," and if he is constrained to put them to the question, to torment them in order to extract their secrets, he is grieved to have provoked "such miseries!" (7/14.)

Having provided for their needs, and satisfied with the secrets which they have revealed to him, it is not without regret and difficulty that he parts from them and restores them "to the delights of liberty."

He is thoroughly convinced, moreover, that all the creatures that share the face of the earth with us are accomplishing an august and appointed task. He welcomes the swallows to his dwelling, even surrendering his workroom to them, at the risk of jeopardizing his notes and books. He pleads for the frog, and applies himself to setting forth his unknown qualities; he rehabilitates the bat, the hedgehog, and the screech-owl, persecuted, defamed, crushed, stoned, and crucified! (7/15.)

So intimate is the life which he leads among them all that he makes himself truly their companion, and relates his own history in narrating theirs; pleased to discover in their joys and sorrows his own trials and delights; mingling in their annals his memories and his impressions; delightful fragments of a childlike autobiography, encrusted in his learned work; moving and delightful pages in which all the ingenuity of this noble mind reveals itself with a touching sincerity, in which all the freshness of this charming and so profoundly unworldly nature is seen as through a pure crystal.

CHAPTER 7.

There is no real communion with nature without sentiment, without an illuminating passion: often the sole and effectual grace which enables its true meaning to appear. Neither taste, nor intelligence, nor logic, nor all the science of the schools can suffice alone. To see further there is needed something like a gift of correspondence, surpassing the limits of observation and experience, which enables us to foresee and to divine the profound secrets of life which lie beneath appearances. Those who are so gifted have often only to open their eyes in order to grasp matters in their true light.

A great observer is in reality a poet who imagines and creates. The microscope, the magnifying glass, the scalpel, are as it were the strings of a lyre. "The felicitous and fruitful hypothesis which constitutes scientific invention is a gift of sentiment" in the words of Claude Bernard; and of this king of physiology, who commenced by proving himself in works of pure imagination, and whose genius finally took for its theme the manifold variations of living flesh, of him too may we not say that he has explored the labyrinths of life with "the torch of poetry in his hand"?

Similarly, do not the harmonious sequences which run through all the admirable discoveries of Pasteur give us the sensation of a veritable and gigantic poem?

In Fabre also it seems that the passion which he brings to all his patient observations is in itself truly creative: "his heart beats with emotion, the sweat drips from his brow to the soil, making mortar of the dust"; he forgets food and drink, and "thus passes hours of oblivion in the happiness of learning." I have seen him in his laboratory studying the spawning of the bluebottle, when I, at his side, could scarcely support the horrible stench which rose from the putrefying adders and lumps of meat; he, however, was oblivious of the frightful odour, and his face was inundated with smiles of delight.

Intelligence, then, must here be the servant of feeling and intuition; a kind of primitive faculty, mysterious and instinctive, which alone makes a great naturalist like Fabre, a great historian like Michelet, a great physician like Boerhaave or Bretonneau.

These last are not always the most scholarly nor the most learned nor the most patient, but they are those who possess in a high degree that special vision, that gift, properly speaking poetic, which is known as the clinical eye, which at the first glance perceives and confirms the diagnosis in all its detail.

Fabre has a mind propitious to such processes; and if, by chance, circumstances had directed his attention to medicine, that science which is based upon an abundant provision of facts, but in which good sense and a kind of divination play a still wider part, there is no doubt that he would have been capable of becoming a shining light in this new arena.

He was full of admiration for that other illustrious Vauclusian, François Raspail (7/16.), whose medical genius anticipated Pasteur and all the conceptions of modern medicine. It would seem that he found in him his own temper, his own fashion of seeing and representing things. He loved Raspail's books and his prescriptions, full of reason and a most judicious good sense, distrusting for himself and for his family the complicated formulae and cunning remedies of an art too considered and still unproved. At Carpentras, while his first-born, Émile, was hovering between life and death, and the physician who came to see him, "being at the end of his resources," did nothing more for him and soon ceased to come, thinking that the child would not last till the morrow, Fabre flew to the works of Raspail.

"I searched to discover what his malady was. I found it, and he was treated day and night accordingly. To-day he is convalescent; and his appetite has returned. I believe he is saved, and I shall say, like Ambroise Paré, 'I have nursed him; God has cured him.'" (7/17.)

The episode which he relates, when, at the primary school of Avignon, a retort had just burst, "spurting in all directions its contents of vitriol," right in the midst of the suddenly interrupted chemistry lesson, and when, thanks to his prompt action, he saved the sight of one of his comrades, does honour to his initiative and presence of mind. (7/18.)

While "all physicians should bow before the facts which he excels in discovering" (7/19.), he has also been able to make direct application of the marvels of entomology to some of the problems of hygiene and medicine. He has shown that the irritant poison secreted by certain caterpillars, "which sets the fingers which handle them on fire," is nothing but a waste product of the organism, a derivative of uric acid; he does not hesitate to perform painful experiments on himself in order to furnish the proof of his theory; and he explains thus the curious cases of dermatitis which are often observed among silkworm-breeders. (7/20.) He proves the uselessness of our meat-safes of metallic gauze, intended to preserve meat against contamination, and the efficacy of a mere envelope of paper, not only to preserve meat from flies, but also our garments from the clothes-moth. (7/21.) He recommends the curious Provençal recipe, which consists in boiling suspected mushrooms in salt and water before eating them. Finally he suggests to members of the medical profession that they might perhaps extract heroic remedies from these treacherous vegetables. (7/22.)

He had need of that indefinite leisure which had hitherto been so wholly lacking, for the events of ephemeral lives occur at indeterminate hours, at unexpected moments, and are of brief duration.

So, attentive to their least movements, Fabre goes forth to observe them at the earliest break of day, in the red dawn, when the bee "pops her head out of her attic window to see what the weather is," and the spiders of the thickets lie in wait under the whorls of their nets, "which the tears of night have changed into chaplets of dewdrops, whose magic jewellery, sparkling in the sun," is already attracting moths and midges.

Seated for hours before a sprig of terebinth, his eye, armed with the magnifying glass, follows the slow manoeuvres of the terebinth louse, whose proboscis "cunningly distils the venom which causes the leaf to swell and produces those enormous tumours, those misshapen and monstrous galls, in which the young pass their period of slumber."

He watches at night, by the dim light of a lantern, to copy the Scolopendra at her task, seeking to surprise the secret of her eggs (7/23.); to observe the Cione constructing her capsule of goldbeater's skin, or the Processional caterpillars travelling head to tail along their satin trail, extinguishing his candle only when sleep at last sets his eyelids blinking. He will wake early to witness the fairy-like resurrection of the silkworm moth (7/24.); "in order not to lose the moment when the nymph bursts her swaddling-bands," or when the wing of the locust issues from its sheath and "commences to sprout"; no spectacle in the world is more wonderful than the sight of "this extraordinary anatomy in process of formation," the unrolling of these "bundles of tissue, cunningly folded and reduced to the smallest possible compass" in the insignificant alar stumps, which gradually unfold "like an immense set of sails," like the "body-linen of the princess" of the fairy-tale, which was contained in one single hemp-seed. (7/25.)

In his Harmas he is like a stranger discovering an unknown world; "like a kindly giant from Sirius, holding a magnifying glass to his eye, retaining his breath, lest it should overturn and sweep away the pigmies which he is observing."

His passion for interrogating the Sphinx of life, everywhere and at all moments, sufficed to fill his days from one end of the year to the other. When some distant subject interested him, even on the most scorching days, he would put "his lunch in his pocket, an apple and a crust of bread," and sit out in the hot sunlight, accompanied by his dog, Vasco,

Tom, or Rabbit; fearing only that some importunate third person might come between nature and himself.

When he walked in his garden he would let nothing escape him; witness those precise notes of an eclipse of the sun, and of the effects which that phenomenon produces upon animal life as a whole.

While his children followed the progress of the moon across the sun through a pane of smoked glass, he attentively observed all that occurred in the countryside.

"It is four; the day grows pale; the temperature is fresher; the cocks crow, surprised by this kind of twilight which comes before the hour. A few dogs are baying...The swallows, numerous before, have all disappeared...a couple have taken refuge in my study, one window of which is open...when the normal light returns they will come outdoors once more...The nightingale, which had so long importuned me by his interminable song, is silent at last (7/26.); the black-capped skylarks, which were warbling continually, are suddenly still...only the young house-sparrows under the tiles of the roof are mournfully chirping...Peace and silence, the daylight more than half gone...In the Harmas I can no longer see the insects flying; I find only one bee pillaging the rosemary; all life has disappeared.

CHAPTER 7.

"Only a weevil, the Lixus," which he is observing in a cage, "continues, step by step, without the slightest emotion, his amorous by-play, as though nothing unusual were happening...The nightingale and the skylark may be silent, oppressed by fear; the bee may re-enter her hive; but is a weevil to be upset because the sun threatens to go out?" (7/27.)

He was no less curious concerning the resurrection of the sun, and every time he made an excursion to the Ventoux he was careful not to miss this spectacle; setting out at an early hour from the foot of the mountain, so that he might see the dawn grow bright from the summit of its rocky mass; then the sun, suddenly rising in the morning breeze, and setting fire, little by little, to the Alps of Dauphiné and the hills of Comtat; and the Rhône, far below, slender as a silver thread.

He took infinite pleasure too in drinking his fill of the sublime terrors of the thunderstorm, which he regarded as one of the most magnificent spectacles which nature can offer; not content with observing it through glass, he would open wide the windows at night the better to enjoy the phosphorescence of the atmosphere, the conflagration of the clouds, the bursts of thunder, and all the solemn pomp with which the great purifying phenomenon manifests itself.

But pure observation, as practised by his predecessors, Réaumur and Huber, is often insufficient, or "furnishes only a glimpse of matters."

He had recourse, therefore, to artificial observation of the kind known as experimentation, and we may say that Fabre was really the first to employ the experimental method in the study of the minds of animals.

Near the field of observation, therefore, is the naturalist's workshop, "the animal laboratory," in which such inductions as may be suggested by the doings and the movements of the insects "which roam at liberty amidst the thyme and lavender" are subjected to the test of experiment. It is a great, silent, isolated room, brilliantly lighted by two windows facing south, upon the garden, one at least of which is always kept open that the insects may come and go at liberty.

In the glass-topped boxes of pine which occupy almost the entire height of the whitewashed walls are carefully arranged the collections so patiently amassed; all the entomological fauna of the South of France, and the sea-shells of the Mediterranean; an abundant wealth also of divers rarities; numismatical treasures and fragments of pottery and other prehistorical documents, of which the numerous ossuaries in the neighbourhood of Sérignan, scattered here and there upon the hills, contain many specimens.

At the top, crowning the facade of glass-topped cases like an immense frieze, is the colossal herbarium, the first volumes of which go back to the early youth of their owner; all the flora, both of the Midi and the North, those of the plains and those of the mountains, and all the algae of fresh and salt water.

But it must not be supposed that Fabre attaches any great value to these collections, enormous though the sum of labour which they represent. To him they have been a means of education, a means of organizing and arranging his knowledge, and not of satisfying an idle curiosity; not the amusement of one content with the rind of things. In order to identify at first sight such specimens as one encounters and proposes to examine, one must first of all learn to observe and to see thoroughly, and to school the eyes in the colours and forms peculiar to each individual species.

One may fairly complain of Réaumur, for example, that his knowledge was uncertain and incomplete. Too often he leaves his readers undecided as to the nature of the species whose habits he describes. Fabre himself, by dint of criticizing with so much humour the abuse of classifications, has sometimes allowed himself to fall into the same fault. (7/28.) He has taken good care, however, not to neglect the systematic study of species; witness his "Flora of the Vaucluse" and that careful catalogue of Avignon which he has not disdained to republish. (7/29.) The truth is that "if we do not know their names the knowledge of the things escapes us" (7/30.), and he was profoundly conscious of the truth of this precept of the great Linnaeus.

The middle of the room is entirely occupied by a great table of walnut- wood, on which are arranged bottles, test-tubes, and old sardine-boxes, which Fabre employs in order to watch the evolution of a thousand nameless or doubtful eggs, to observe the labours of their larvae, the creation and the hatching of cocoons, and the little miracles of metamorphosis, "after a germination more wonderful than that of the acorn which makes the oak."

Covers of metallic gauze resting on earthenware saucers full of sand, a few carboys and flower-pots or sweetmeat jars closed with a square of glass; these serve as observation or experimental cages in which the progress and the actions of "these tiny living machines" can be examined.

Fabre has revealed himself as a psychologist without rival, of a consummate skill in the difficult and delicate art of experimentation; the art of making the insect speak, of putting questions to it, of forcing it to betray its secrets; for experiment is "the only method which can throw any light upon the nature of instincts."

His resources being slender and his mind inventive, he has ingeniously supplemented the poverty of his equipment, and has discovered less costly and less complex means of conducting his experiments; knowing the secret of extracting the sublimest truth from clumsy combinations of "trivial, peasant-made articles."

He has succeeded, in his rustic laboratory, in applying the rigorous rules of investigation and experimentation established by the great biologists. He has therefore been able to establish his beautiful observations in a manner so indisputable that those who come after him and are tempted to study the same things can but arrive at the same results, and derive inspiration from his researches.

To note with care all the details of a phenomenon is the first essential, so that others may afterwards refer to them and profit by them; the difficult thing is to interpret them, to discover the circumstances, the whys and wherefores, the consequences, and the connecting links.

But a single fact observed by chance at the wayside, and which would not even attract the attention of another, will be instantly luminous to this searching understanding, it will suggest questions unforeseen, and will evoke, by anticipation, preconceived ideas and sudden flashes of intuition, which will necessitate the test of experiment.

Why, for example, does the Philanthus, that slender wasp, which captures the honey-bee upon the blossoms in order to feed her larvae; why, before she carries her prey to her offspring, does she "outrage the dying insect," by squeezing its crop in order to empty it of honey, in which she appears to delight, and does indeed actually delight?

"The bandit greedily takes in her mouth the extended and sugared tongue of the dead insect; then once more she presses the neck and the thorax, and once more applies the pressure of her abdomen to the honey-sac of the bee. The honey oozes forth and is instantly licked up. Thus the bee is gradually compelled to disgorge the contents of the crop. This atrocious meal lasts often half an hour and longer, until the last trace of honey has disappeared."

The detailed answer is obtained by experiment, which perfectly explains this "odious feast," the excuse for which is simply maternity. The Philanthus knows, instinctively, without having learned it, that honey, which is her ordinary fare, is, by a very singular "inversion," a mortal poison to her larvae. (7/31.)

As an accomplished physiologist, Fabre conducts all kinds of experiments. Behind the wires of his cages, he provokes the moving spectacle of the scorpion at grip with the whole entomological fauna, in order to test the effects of its terrible venom upon various species; and thus he discovers the strange immunity of larvae; the virus, "the reagent of a transcendent chemistry, distinguishes the flesh of the larva from that of the adult; it is harmless to the former, but mortal to the latter"; a fresh proof that "metamorphosis modifies the substance of the organism to the point of changing its most intimate properties." (7/32.)

You may judge from this that he knows through and through the history of the creatures which form the subjects of his faithful narratives. He is informed of the smallest events of their lives. He possesses a calendar of their births; he records their chronology and the succession of generations; he has noted their methods of work, examined their diet, and recorded their meals. He discovers the motives which dictate their peculiarities of choice; why the Cerceris, for instance, among all the victims at its disposal, never selects anything but the Buprestis and the weevils. He is familiar too with their tactics of warfare and their methods of conflict.

CHAPTER 7.

His gaze has penetrated even the most hidden dwellings; those in which the Halictus "varnishes her cells and makes the round loaf which is to receive the egg"; in which, under the cover of cocoons, murderous grubs devour slumbering nymphs; even the depths of the soil are not hidden from him, for there, thanks to his artifices, he has surprised the astonishing secret of the Minotaur.

He sifts all doubtful stories; anecdotes, statements of supposed habits; all that is incoherent, or ill observed, or misinterpreted; all the cliches which the makers of books pass from hand to hand.

In place of repetition he gives us laws, constant facts, fixed rules.

With incomparable skill, he repeats and tests the ancient experiments of Réaumur.

He is not content to show us that Erasmus Darwin is mistaken; he points out how it is that he has fallen into error. (7/33.)

He sets himself to decipher the meaning of old tales, skilfully disengaging the little parcel of truth which usually lies beneath a mass of incorrect or even false statements. He criticises La Fontaine, and questions the statements of

Horus Apollo and Pliny. From a mass of undigested knowledge he has created the living science of entomology, which had received from Réaumur a first breath of vitality, in such wise that each individual creature is presented in his work with its precise expression and the absolute truth of its character and attitudes; the inhabitants of the woods and fields, whether those which feed upon the crops or those which live in the crevices of the rocks, or the obscure workers that crawl upon the earth; all those which have a secret to tell or something to teach us; the Cigale, so different from the insect of the Fable; and above all that beetle whose name had hitherto been encountered arrayed in the most fantastic legends, the famous Scarabaeus sacer of the tombs, which Fabre preferred to place at the head of his epic as an agreeable prologue, although the inquiry relative to his amazing feats belongs chronologically to a comparatively recent period of his career.

How moderate he is in such suppositions as he ventures; how cautious when his persistent patience has at last struck against "the inaccessible wall of the Unknowable"! Then, with admirable frankness, tranquil and sincere, he simply owns that "he does not know," unlike so many others, whose uncritical minds are contented with a fragmentary vision, and run so far ahead of the facts that they can only promote indefinite illusion and error.

CHAPTER 7.

One is surprised indeed to remark how few even of the most learned and well-informed of men have a real aptitude for observation, and a highly instructive book might be written concerning the discrepancies and the weak points in our knowledge. If they were subjected to a sufficiently severe test, how threadbare would appear many of those problems which nature and the world present, and which are regarded as resolved!

How long, for instance, was needed to destroy the legend of the cuckoo, incessantly repeated down to the days of Xavier Raspail, and to us so familiar; to elucidate its history, and to set it in its true light! (7/34.)

It is by means of such data as these that a science is founded, for theories decay, and only well-observed facts remain irrefragable. With stones such as these, which are hewn by the great artisan, the structures of the future will be built, and our own science, perhaps, will one day be refashioned.

For this reason Fabre's books are an education for all those who wish to devote themselves to observation; a manual of mental discipline, a true "essay upon method," which should be read by every naturalist, and the most interesting, instructive, familiar and delightful course of training that has ever been known.

On the other hand, it is impossible to conceive what labour this delicate work demands; what perseverance Fabre has required painfully to extract one grain of gold; to glean and unite the definite factors, the positive documents, which served as foundations for each of his essays; lucid, limpid, and captivating as the most delightful of fairy-tales. We are charmed, fascinated, and astonished; we see nothing of the groping advance, the checks, and all the toil and the patience demanded. We do not suspect the long waiting, the hesitation, the desperate length of the inquiries. For example, to establish the curious relations which exist between the wasps and the Volucellae, what long and repeated experiments were needful! His notebooks, in which he records, from day to day, all that he sees, are evidence of this. What watches in the alley of lilacs, year after year, to decipher the mechanism and the mode of construction of the hunting-net of the Epeïra! Some of these histories, like that of the hyper-metamorphosis of the Meloë, were only completed as the result of twenty-five years of assiduous inquiry, while forty years were required to complete that of the Scarabaeus sacer, for his observation of it was always partial; it is almost always impossible to divine what one cannot see from the little that one does see; and as a rule one must return to the same point over and over again in order to fill up lacunae.

CHAPTER 7.

The majority of the insects which Fabre has studied are solitary, and are only to be encountered singly, scattered over wide areas of country. Some live only in determined spots, and not elsewhere, such as the famous Cerceris, or the yellow-winged Sphex, of which no trace is to be found beyond the limits of the Carpentras countryside.

The proper season must be watched for; one must be ready at any moment to profit by a lucky chance, and resign oneself to interminable watches at the bottom of a ravine, or keep on the alert for hours under a fiery sun. Often the chance goes by, or the trail followed proves false; but the season is over, and one must wait for the return of another spring. The trade of observer in many cases resembles the exhausting labours of the Sisyphus beetle, painfully pushing his pellet up a rough and stony path; so that the team halts and staggers at every moment, the load spills over and rolls away, and all has to be commenced over again.

We can now cast back, in order to consider at leisure the immortal study which marked the beginning of his fame, with the greater interest and profit in that Fabre has been able, during his retirement, to generalize and extend his discovery. (7/35.)

CHAPTER 7.

Let us first of all note how the observation which Dufour had made of the nest of the Cerceris was transformed in his hands, and what developments he was able to evolve therefrom.

Since they have been definitely established by Fabre these curious facts have been well-known. They form perhaps the greatest prodigy presented by entomology, that science so full of marvels.

These wasps nourish themselves only on the nectar of flowers; but their larvae, which they will never behold, must have fresh and succulent flesh still palpitating with life.

The insect digs a tunnel in the soil, in which she places her eggs, and having provisioned the cell with selected game--cricket, spider, caterpillar, or beetle--she finally closes the entrance, which she does not again cross.

Like nearly all insects, the young wasp is born in the larval state, and from the moment of its hatching to the end of its growth--that is to say, for a period of many days--the grub enclosed in its cell can look for no help from without.

Here then is a fascinating problem: either the victims deposited by the mother are dead, and desiccation or putrefaction attacks them promptly, or else they are living, as indeed the larvae require; but then "what will become of this fragile creature, which a mere nothing will destroy, shut in the narrow chamber of the burrow among vigorous beetles, for weeks on end working their long spurred legs; or at grips with a monstrous caterpillar making play with its flanks and mandibles, rolling and unrolling its tortuous folds?"

Such is the thrilling mystery of which Fabre discovered the key.

With inconceivable ingenuity, the victim is seized and thrown to the ground, and the wasp plunges her sting, not at random into the body, which would involve the risk of death, but at determined points, exactly into the seat of those invisible nervous ganglions whose mechanism commands the various movements of the creature.

Immediately after these subtle wounds the prey is paralysed throughout its body; its members appear to be disarticulated, "as though all the springs were broken"; the true corpse is not more motionless.

But the wound is not mortal; not only does the insect continue to live, but it has acquired the strange prerogative of being able to live for a very long period without taking any nourishment, thanks precisely to the condition of immobility, in some sort vegetative, which paralysis confers upon it.

When the hour strikes the hungry larva will find its favourite meat served to its liking; and it will attack this defenceless prey with all the circumspection of a refined eater; "with an exquisitely delicate art, nibbling the viscera of its victim little by little, with an infallible method; the less essential parts first of all, and only in the last instance those which are necessary to life. Here then is an incomprehensible spectacle; the spectacle of an animal which, eaten alive, mouthful by mouthful, during nearly a fortnight, is hollowed out, grows less and less, and finally collapses," while retaining to the end its succulence and its freshness.

The fact is that the mother has taken care to deposit her egg "at a point always the same" in the region which her sting has rendered insensible, so that the first mouthfuls are only feebly resented. But as the enemy goes deeper and deeper "it sometimes happens that the cricket, bitten to the quick, attempts to retaliate; but it only succeeds in opening and closing the pincers of its mandibles on the empty air, or in uselessly waving its antennae." Vain efforts: "for now the voracious beast has bitten deep into the spot, and can with impunity ransack the entrails." What a slow and horrible agony for the paralysed victim, should some glimmer of consciousness still linger in its puny brain!

What a terrible nightmare for the little field-cricket, suddenly plunged into the den of the Sphex, so far from the sunlit tuft of thyme which sheltered its retreat!

To paralyse without killing, "to deliver the prey to the larvae inert but living": that is the end to be attained; only the method varies according to the species of the hunter and the structure of the prey; thus the Cerceris, which attacks the coleoptera, and the Scolia, which preys upon the larvae of the rose-beetle, sting them only once and in a single place, because there is concentrated the mass of the motor ganglions.

The Pompilus, which selects a spider for its victim, no less than the redoubtable Tarantula, knows that its quarry "has two nervous centres which animate respectively the movements of the limbs and those of the terrible fangs; hence the two stabs of the sting." (7/36.)

The Sphex plunges her dagger three times into the breast of the cricket, because she knows, by an intuition that we cannot comprehend, that the locomotor innervation of the cricket is actuated by three nervous centres, which lie wide apart. (7/37.)

Finally, the Ammophila, "the highest manifestation of the logic of instinct, whose profound knowledge leaves us confounded, stabs the caterpillar in nine places, because the body of the victim with which it feeds its larvae is a series of rings, set end to end, each of which possesses its little independent nervous centre." (7/38.)

This is not all; the genius of the Sphex is not yet at the end of its foresight. You have doubtless heard of the comatose state into which the wounded fall when, after a fracture of the skull, the brain is compressed by a violent haemorrhage or a bony splinter. The physiologists imitate this process of nature when they wish, for example, to obtain, in animals under experiment, a state of complete immobility. But did the first surgeon who thought of trepanning the skull in order to exert on the brain, by means of a sponge, a certain degree of compression, ever imagine that an analogous procedure had long been employed in the insect world, and that these clumsy methods were merely child's play beside the astonishing feats of the Unconscious?

For the stab in the thoracic ganglions, however efficacious, is often insufficient. Although the six limbs are paralysed, although the victim cannot move, its mandibles, "pointed, sharp, serrated, which close like a pair of scissors, still remain a menace to the tyrant; they might at least, by gripping the surrounding grasses, oppose a more or less effectual resistance to the process of carrying off." So the

CHAPTER 7. 176

preceding manoeuvres are consummated by a kind of garrotting; that is, the insect "takes care to compress the brain of its victim, but so as to avoid wounding it; producing only a stupor, a simple torpor, a passing lethargy." Is not the ingenious observer justified in concluding that "this is alarmingly scientific"?

Between the dry statements of Dufour, which served Fabre as his original theme, and the unaccustomed wealth of this vast physiological poetry, what a distance has been covered!

How far have we outstripped this barren matter, these shapeless sketches! Dufour, another solitary, who retired to his province, in the depth of the Landes, was above all a descriptive anatomist, and he limited himself to an inventory of the nest of a Cerceris.

For him the Buprestes were dead, and their state of preservation was explained simply as a kind of embalming, due to some special action of the venom of the Hymenoptera.

These facts, therefore, were stated as simple curiosities.

Fabre proved that these victims possessed all the attributes of life excepting movement, by provoking contractions in their members under the influence of various stimulants, and by keeping them alive artificially for an indefinite period.

On the other hand, he demonstrated the comparative innocuousness of the venom of these wasps, some of which, like the great Cerceris or the beautiful and formidable Scolia, alarm by their enormous size and their terrifying aspect; so that the conservation of the prey could not be due to any occult quality, to some more or less active antiseptic virtue of the venomous fluid, but simply to the precision of the stab and the miraculous deftness of the "surgeon."

He also pointed out the fact that the sting of the insect is able immediately to dissociate the nervous system of the vegetative life from that of the correlative life, sparing the former, and taking care not to wound the abdomen, which contains the ganglions of the great sympathetic nerve, while it annihilates the latter, which is more or less concentrated along the ventral face of the thoracic region.

He completed this splendid demonstration, not only by provoking under his own eyes the "murderous manoeuvres, the intimate and passionate drama," but also by reproducing experimentally all these astonishing phenomena; expounding their mechanism and their variations with a logic and lucidity, an art and sagacity which raise this marvellous observation, one of the most beautiful known to science, to the height of the most immortal discoveries of physiology. Claude Bernard, in his celebrated experiments, certainly exhibited no greater invention, no truer genius.

CHAPTER 8.

THE MIRACLE OF INSTINCT.

"The Spirit Bloweth Whither it Listeth."

What is this instinct, which guides the insect to such marvellous results? Is it merely a degree of intelligence, or some absolutely different form of activity?

Is it possible, by studying the habits of animals, to discover some of those elementary springs of action whose knowledge would enable us to dive more deeply into our own natures?

Fabre has presented us to his Sphex, the "infallible paralyser." Are we to credit her not only with memory, but also with the faculty of associating ideas, of judgment, and of pursuing a train of reasoning in respect of her astonishingly co-ordinated actions?

Put to the question by the malice of the operator, the "transcendent" anatomist trips over a mere trifle, and the slightest novelty confounds her.

Without the circle of her ordinary habits, what stupidity, "what darkness wraps her round"! She retreats; she refuses to understand; "she washes her eyes, first passing her hands across her mouth; she assumes a dreamy, meditative air." What can she be pondering? Under what form of thought, illusion, or mirage does the unfamiliar problem which has obtruded itself into her customary life present itself behind those faceted eyes? (8/1.)

How can we tell? We can only attain to knowledge of ourselves by direct intuition. It is only the idea of our ego which enables us to conjecture what is passing in the brains of our fellows. Between the insect and ourselves no understanding is possible, so remote are the analogies between its organization and our own; and we can only form idle hypotheses as to its states of consciousness and the real motive of its actions.

Consider only that unknown and mysterious energy which the insects display in their operations and their labours, as it is in itself, and let us content ourselves, first of all, with comparing it to our own intelligence, such as we conceive it to be.

In seeking to appreciate whereby it differs perhaps we shall gain more than by vainly seeking points of resemblance. We shall discover, in fact, behind the insect and its prodigious instincts, a vast and remote horizon, a region at once more profound, more extensive, and more fruitful

CHAPTER 8.

than that of the intelligence; and if Fabre is able to help us to decipher a few pages of "the most difficult of all volumes, the book of ourselves," it is precisely, as a philosopher told him, because "man has remained instinctive in process of becoming intelligent." (8/2.)

The work of Fabre is from this point of view an invaluable treasury of observations and experiments, and the richest contribution which has ever been made to the study of these fascinating problems.

"The function of the intelligence is to reflect, to be conscious; that is, to relate the effect to its cause, to add a "because" to a "why"; to remedy the accidental; to adapt a new course of conduct to new circumstances."

In relation to the human intelligence thus defined Fabre has considered these nervous aptitudes, so well adjusted, according to the evolutionists, by ancient habit, that they have finally become impulsive and unconscious, and, properly speaking, innate. He has demonstrated, with an abundance of proof and a power of argument that we must admire, the blind mechanism which determines all the manifestations, even the most extraordinary, of that which we call instinct, and which heredity has fixed in a species of unchangeable automatism, like the rhythm of the heart and the lungs. (8/3.)

CHAPTER 8.

Let us, from this wealth of material, from among the most suggestive examples, select some of his most striking demonstrations, which are classics of their kind.

Fabre has not attempted to define instinct, for it is indefinable; nor to probe its essential nature, which is impenetrable. But to recognize the order of nature is in itself a sufficiently fascinating study, without striving to crack an unbreakable bone or wasting time in pondering insoluble enigmas. The important matter is to avoid the introduction of illusions, to beware of exceeding the data of observation and experiment, of substituting our own inferences for the facts, of outstripping reality and amplifying the marvellous.

Let us listen to the scrupulous analysis whose lessons, scattered through four thousand pages, teach us more concerning instinct and its innumerable variations than all the most learned treatises and speculations of the philosophers.

Nothing in the world perplexes the mind of the observer like the spectacle of the birth and growth of the instincts.

At precisely the right moment, just as failure or disaster seems foreordained by the previously established circumstances, Fabre shows us his insects as suddenly mastered by an irresistible force.

"At the right moment" they invincibly obey some sort of mysterious and inflexible prescription. Without apprenticeship, they perform the very actions required, and blindly accomplish their destiny.

Then, the moment having passed, the instincts "disappear and do not reawaken. A few days more or less modify the talents, and what the young insect knew the adult has often forgotten." (8/4.)

Among the Lycosae, at the moment of exodus, a sudden instinct is evolved which a few hours later disappears never to return. It is the climbing instinct, unknown to the adult spider, and soon forgotten by the emancipated young, who are destined to roam upon the face of the earth. But the young Lycosae, anxious to leave the maternal home and to travel, become suddenly ardent climbers and aeronauts, each releasing a long, light thread which serves it as parachute. The voyage accomplished, no trace of this ingenuity is left. Suddenly acquired, the climbing instinct no less suddenly disappears. (8/5.)

The great historiographer of instinct has thrown a wonderful light, by his beautiful experiments relating to the nidification of the mason-bee, upon the indissoluble succession of its different phases; the lineal concatenation, the inevitable and necessary order which presides over each of these nervous discharges of which the total series constitutes, properly speaking, a mode of action.

The mason-bee continues to build upon the ready-completed nest presented to her. She obstinately insists upon provisioning a cell already duly filled with the quantity of honey required by the larva, because, in this case as in the other, the impulse which incites her to build or to provision the nest has not yet been exhausted.

On the other hand, if we empty the little cup of its contents when she has filled it she will not recommence her labours. "The process of provisioning being complete, the secret impulse which urged her to collect her honey is no longer active. The insect therefore ceases to store her honey, and, in spite of this accident, lays her egg in the empty cell, thus leaving the future nursling without nourishment." (8/6.)

In the case of the Pelopaeus, Fabre calls our attention to one of the most instructive physiological spectacles that can be imagined.

While the mason-bee does not notice that her cell has been emptied, the Pelopaeus cannot perceive that the tricks of the experimenter have resulted in the

CHAPTER 8.

185

disappearance of her progeny; and she "continues to store away spiders for a germ that no longer exists; she perseveres untiringly in her useless hunting, as though the future of her larva depended on it; she amasses provisions which will feed no one; more, she pushes aberration to the extent of plastering even the place where her nest was if we remove it, giving the last strokes of the trowel to an imaginary building, and putting her seals upon empty nothing." (8/7.)

>From these facts, and others, no less celebrated, which show "the inability of insects to escape from the routine of their customs and their habitual labours," Fabre derives so many proofs of their lack of intelligence.

The Epeïra fasciata is incapable of replacing a single radial thread in the geometrical structure of its web, when broken; it recommences the entire web every evening, and weaves it at one stretch with the most beautiful mastery, as though merely amusing itself.

The caterpillar of the Greater Peacock moth teaches us the same lesson; when occupied in weaving its cocoon it does not know how to repair an artificial rent; and "in spite of the certainty of its death, or rather that of the future butterfly, it quietly continues to spin, without troubling to cover the rent; devoting itself to a superfluous task, and ignoring the treacherous breach, which leaves the cocoon and its inhabitant at the mercy of the first thief that finds it." (8/8.)

Thus "because one action has just been performed, another must inevitably be performed to complete the first; what is done is done, and is never repeated. Like the watercourse, which cannot climb the hills and return to its source, the insect does not retrace its steps or repeat its actions, which follow one another invariably, and are inevitably connected in a necessary order, like a series of echoes, one of which awakens another...The insect knows nothing of its marvellous talents, just as the stomach knows nothing of its cunning chemistry. It builds like a bricklayer, weaves, hunts, stabs, and paralyses, as it secretes the venom of its weapons, the silk of its cocoon, the wax of its comb, or the threads of its web; always without the slightest knowledge of the means and the end." (8/9.)

Thus instinct is one thing and intelligence is another; and for Fabre there is no transition which can transform the one into the other.

But how profound and abundant, how infinite is the source from which this manifold activity derives, distributed as it is throughout the entire animal kingdom; and which in ourselves commands the profoundest part of our nature;

unconscious, or even in opposition to our wonderful intelligence, which it often silences or altogether overwhelms.

Although the insect "has no need of lessons from its elders" in order to accomplish its beautiful masterpieces, the comprehensive concept of the genius which rises spontaneously and at a single step to the loftiest conceptions is not always a product of pure reason.

Compare the sublime logic of animal maternity, the impeccable dictates of instinct, with the hesitations, the gropings, the uncertainties, the errors and tragic failures of human maternity, when it seeks to replace the unerring commands of instinct by the clumsy efforts of the intelligence!

If all is darkness to the animal, apart from its habitual paths, how feeble and hesitating, how faltering and unequal is reason when it seeks to oppose its laborious inductions to the infallible wisdom of the unconscious!

It is, in fact, to this concatenation of actions, narrowly connected by a mutual dependence, that we owe this inexhaustible series of cunning industries and wonderful arts. To Fabre they are so many feats of a learned unconsciousness.

"See the nest, the accustomed masterpiece of mothers; it is more often than otherwise an animal fruit, a coffer full of germs, containing eggs in place of seeds."

The satin bag of the Epeïra fasciata, in which her eggs are enclosed, "breaks at the caress of the sun, like the skin of an over-ripe pomegranate."

The Dorthesia, the louse inhabiting the euphorbia, "trebles the length of her body, prolonging its hinder part into a pouch, comparable to that of the opossum, into which the eggs are dropped, and in which the young are hatched, to leave it afterwards at will." (8/10.)

The Chermes of the ilex "hardens into a rampart of ebony, whence an innumerable legion of vermin bursts forth one day without changing their place."

The capsule of gold-beater's skin, in which the grubs of the Cione are enclosed, divides itself, at the moment of liberation, into two hemispheres "of a regularity so perfect that they recall exactly the bursting of the pyxidium when the seed is distributed." (8/11.)

Here and there, however, we catch a glimpse of a rudiment of what we understand by consciousness, in the shape of a "vague discrimination."

CHAPTER 8.

Each plant has its lover, drawn to it by a kind of elective affinity and invariable tendency. The Larra makes for the thistle, the Vanessa for the nettle, the Clytus for the ilex, and the Crioceris for the lily. "The weevil knows nothing but its peas and beans, the golden Rhynchites only the sloe, and the Balaninus only the nut or acorn."

But the Pieris, which haunts the cabbage, frequents the nasturtium also, and the golden rose-beetle, which "intoxicates itself at the clusters of the hawthorn," is no less addicted to the nectar of the rose.

The Xylocopa, which burrows in the trunks of trees and old rafters, forming little round corridors in which to lodge her offspring, "will utilize artificial galleries which she has not herself bored."

The Chalicodoma "also is aware of the economic advantages of an old abandoned nest"; the Anthophora is careful to establish her family "at the least expense," and profits on occasion by galleries which have been mined by previous generations; adapting herself to these new conditions, she repairs the tunnels which she did not construct "and economizes her forces." (8/12.)

It would seem, therefore, that these tiny minds are created and shaped by means of experience; they recognize "that which is most fitting"; they learn, they compare; may we not also say that they judge?

Does not the Mason-bee, "which rakes the roads for a dry powdery dust and mixes it with saliva to convert it into a hard cement," foresee that this mud will harden?

Is the Pelopaeus devoid of judgment when she seeks the interior of dwelling-houses in order to shelter her nest of dried clay, which the least drop of rain would reduce to its original state of mud?

Is it without knowledge of the effects that the sloe-weevil builds a ventilating chimney to prevent the asphyxiation of her larva? that the Scarabaeus sacer contrives a filter at the smaller end of its pear-shaped ball, by means of which the grub is able to breathe? or that Arachne labyrintha "introduces in her silk-work a rampart of compressed earth to protect her eggs from the probe of the Ichneumon"?

May we not also see a masterpiece of the highest logic in the house of the trap-door spider, Arachne clotho, which is furnished with a door, a true door "which she throws open with a push of the leg, and carefully bolts behind her on returning by means of a little silk"? (8/13.)

What a miracle of invention too is the prodigious nest of the Eumenes, "with its egg suspended by a thread from the roof, like a pendulum, oscillating at the lightest breath in order to save it from contact with the caterpillars, which, incompletely paralysed, are wriggling and writhing below"! Later, when the egg is hatched, "the filament is transformed into a tube, a place of refuge, up which the grub clambers backwards. At the least sign of danger from the mass of caterpillars the larva retreats into its sheath and ascends to the roof, where the wriggling swarm cannot reach it." (8/14.)

Let us refer also to the remarkable history of the Copris. We cannot deny that the valiant dung-beetle is capable of "evading the accidental" (which to Fabre constitutes one of the distinctive characteristics of the intelligence), since it immediately intervenes if with the point of a penknife we open the roof of its nest and lay bare its egg. "The fragments raised by the knife are immediately brought together and soldered, so that no trace is left of the injury, and all is once more in order." We may read also with what incredible address the mother Copris was able to use and to profit by the ready-made pellets of cow-dung which it occurred to Fabre to offer her. (8/15.)

But their scope is limited, and encroaches very little, in the eyes of the great observer, on the domain of intelligence.

CHAPTER 8.

This he demonstrates to satiety, and his astonishing Necrophori, which adapt themselves so admirably to circumstances and triumph over the experimental difficulties to which he subjects them, seem scarcely to exceed the limits of those actions which at bottom are merely unconscious. (8/16.)

With the spawning of the Osmia, Fabre throws a fresh and unexpected light on the intuitive knowledge of instinct.

We are still groping our way among the causes which rule the determination of the sexes. Biology has only been able to throw a few scattered lights on the subject, and we possess only a few approximate data; which nevertheless are turned to account by the breeders of insects. We are still in the region of illusion and imperfect prognostics.

But the Osmia knows what we do not. She is deeply versed in all physiological and anatomical knowledge, and in the faculty of creating children of either sex at will.

These pretty bees, "with coppery skin and fleece of ruddy velvet," which establish their progeny in the hollow of a bramble stump, the cavity of a reed, or the winding staircase of an empty snail-shell, know the fixed and immutable genetic laws which we can only guess at, and are never mistaken.

CHAPTER 8.

This marvellous prerogative the Osmia shares with a host of apiaries, in which the unequal development of the males and females requires an unequal provision of space and of nourishment for the future larvae. For the females, who exceed in point of size, huge cells and abundant provision; for the more puny males, narrow cells and a smaller ration of pollen and honey.

Now the circumstances which are encountered by the Osmia, when, pressed by the necessities of spawning, she searches for a dwelling, are often fortuitous and incapable of modification; and in order to give each set of larvae the necessary space "she lays at will a male or a female egg, according to the conditions of space."

In this marvellous study, which constitutes, with the history of the Cerceris, the finest masterpiece of experimental entomology, Fabre brilliantly establishes all the details of that curious law which in the Hymenoptera rules both the distribution and the succession of the sexes. In his artificial hives, in glass cylinders, he forces the Osmia to commence her spawning with the males, instead of beginning with the females as nature requires, since the insect is primarily preoccupied with the more important sex, that which ensures par excellence the perpetuation of the species. He even forces the whole swarm which buzzes about his work-tables, his books, his bottles, and apparatus, completely to change the order of its spawning. He shows finally that in the heart of the ovaries the egg of the Osmia has as yet no determined sex, and that it is only at the precise moment when the egg is on the point of emerging from the oviduct that it receives, AT THE WILL OF THE MOTHER, the mysterious, final, and inevitable imprint.

But whence does the Osmia derive this, "distinct idea of the invisible"? Here again is one of those riddles of nature which Fabre declares himself quite incapable of solving. (8/17.)

Is this all? No; we are far from having made the tour of this miraculous and incommensurable kingdom through which this admirable master leads us, and I should never be done were I to attempt to exhaust all the spectacles which he offers us. Let us descend yet another step, among creatures yet smaller and humbler. We shall find tendencies, impulses, preferences, efforts, intentions, "Machiavellic ruses and unheard-of stratagems."

Certain miserable black mites, living specks, the larvae of a beetle, one of the Meloidae, the Sitaris, are parasites of the solitary bee, the Anthophora. They wait patiently all the winter at the entrance of her tunnel, on the slope of a sunny bank, for the springtime emergence of the young bees, as yet imprisoned in their cells of clay. A male Anthophora, hatched a little earlier than the females, appears in the entrance of the tunnel; these mites, which are armed with robust talons, rouse themselves, hasten to and fro, hook themselves to his fleece, and accompany him in all his peregrinations; but they quickly recognize their error; for these animated specks are well aware that the males, occupied all day long in scouring the country and pillaging the flowers, live exclusively out of doors, and would in no wise serve their end. But the moment comes when the Anthophora pays court to the fair sex, and the imperceptible creature immediately profits by the amorous encounter to change its winged courser. "These pigmies therefore have a memory, an experience of facts" (and how one is tempted to add, a glimmering of intelligence!). Grappled now to the female bee, the grub of the Sitaris "conceals itself, and allows itself to be carried by her" to the end of the gallery in which she is now contriving her cradle, "watches the precise moment when the egg is laid, installs itself upon it, and allows itself to fall therewith upon the surface of the honey, in order to substitute itself for the future offspring of the Anthophora, and possess itself of house and victuals." (8/18.)

Another "little gelatinous speck," "a shadow of a creature," the larva of a Chalcidian, the Leucopsis, one of the parasites of the Mason-bee, knows that in the cell of the mason there is food for one only. Scarcely has it entered the tiny dwelling but we see this "nameless shape" for several days "anxiously wandering; it visits the top and bottom, the back, the front, the sides"; it makes the tour of its domain; "it searches in the darkness, palpitating, seemingly with an object in view." What does this "animated globule" want? why is this atom so excited? It is searching to discover if there is not in some corner hitherto unexplored another larva, a rival, that it may exterminate it! (8/19.)

What then intrinsically is instinct? And what intrinsically is intelligence?

How can we propose to draw up the inexhaustible inventory of all the manifestations of life, and why attempt to include all its species and their unknown varieties in narrow classes? Why say that there are only two modes of life, instinct on the one hand and intelligence on the other, "when we know how subtle and illusive is this Proteus, and that there are not two things only, but a thousand dissimilar things" (8/20.): or rather is it not always the same thing, everywhere present and acting in living matter, and susceptible of infinite degrees, under forms and disguises innumerable?

This is why it escapes the "scalpel of the masters" and the apparatus of the chemists. We may dissect, we may scrutinize organs under the magnifying glass, examine wing-cases, count the nervures of the wings, the number of articulations in the limbs; we may reckon every point, like Réaumur forgetting not a line, not a hair; we may compare and measure every portion of the mouth, and define the class; and we shall not find a single point in all this physical architecture which will positively inform us of the habits of the insect. Of what account are a few slight differences? It is in the physical far more than in the anatomical differences that the inviolable demarcation between two species exists. Instincts dominate forms; the tool does not make the artisan; "and none of these various structures, however well adapted they may appear to us, bears within it its reason or its finality."

Thus whatever opinion we may hold as to the nature of instinct, the accomplishments and habits of insects are not, properly speaking, connected with the external and visible form of their organs, and their acts do not necessarily presuppose the instruments which would be appropriate to them.

We know that with most organisms, and particularly with plants, an almost imperceptible variation in material circumstances is often enough to modify their character and to produce fresh aptitudes. Nevertheless, we can but wonder, with Fabre, that physical modifications, which, when they do exist, are so slight always as to have escaped the most perfect observation, should have sufficed to determine the appearance of profoundly dissimilar faculties. Inexplicable abilities, unexpected habits, unforeseen physical aptitudes, and unheard-of industries are exercised by means of organs which are here and there practically identical. "The same tools are equally good for any purpose. Talent alone is able to adapt them to manifold ends."

The Anthidia have two particular industries; "those which felt cotton and card the soft down of hairy plants have the same claws, the same mandibles, composed of the same portions as those which knead resin and mix it with fine gravel." (8/21.)

The sloe-weevil "bores the hard stone of the sloe with the same rostrum as that which its congeners, so like it in conformation, employ to roll the leaves of the vine and the poplar into tiny cigars."

The implement of the Megachile, the rose-fly, is by no means appropriate to its industry; "yet the perfectly circular fragments of leaves have the precise perfection of form that a punch would give."

CHAPTER 8.

The Xylocopa, in order to pierce wood and to bore its galleries in an old rafter, employs "the same utensils which in others are transformed into picks and mattocks to attack clay and gravel, and it is only a predisposition of talent that holds each worker to his speciality."

Moreover, have not the superior animals the same senses and the same structure, yet what inequality there is among them, in the matter of aptitudes and degrees of intelligence!

Habits are no more determined by anatomical peculiarities than are aptitudes or industries.

The two Goat-moth caterpillars, of similar structure, have entirely different stomachic aptitudes; "the exclusive portion of the one is the oak and of the other the hawthorn or the cherry-laurel."

"Whence does the Mantis derive its excessive hunger, its pugnacity, its cannibalism, and the Empusa its sobriety, its peaceableness, when their almost identical organization would seem to indicate an identity of needs, instincts, and habits?"

In the same way the black scorpion appears to present none of the interesting peculiarities which we observe in the habits of its congener, the white scorpion of Languedoc. (8/22.)

Structure, therefore, tells us nothing of aptitude; the organ does not explain its function. Let the specialists hypnotize themselves over their lenses and microscopes; they may accumulate at leisure masses of details relating to this or that family or genus or individual; they may undertake the most subtle inquiries, may write thousands and thousands of pages in order to detail a few slight variations, without even succeeding in exhausting the matter: they will not even have seen what is most wonderful.

When the little insect has for the last time cleaned its claws, the secret of the little mind has fled for ever, with all the feelings that animated it and gave it life. That which is crystallized in death cannot explain what was life. This is the thought which the Provençal singer, with that intuition which is the privilege of genius, has expressed in these melodious lines:

"Oh! pau de sèn qu'emé l'escaupre Furnant la mort, creson de saupre, La vertu de l'abiho e lou secrèt doù méu."

(O men of little sense, who seek, Scalpel in hand, to make Death tell The virtue of the bee, the secret of her cell!) (8/23.)

CHAPTER 9.

EVOLUTION OR "TRANSFORMISM."

"How did a miserable grub acquire its marvellous knowledge? Are its habits, its aptitudes, and its industries the integration of the infinitely little, acquired by successive experiences on the limitless path of time?"

It is in these words that Fabre presents the problem of evolution.

Difficult though it may be to follow the sequence of forms which have endlessly succeeded and replaced one another on the face of the earth, since the beginning of the world, it is certain that all living creatures are closely related; and the magnificent and fertile hypothesis of evolution, which seeks to explain how extant forms are derived from extinct, has the immense advantage of giving a plausible reason for the majority of the facts which at least cease to be completely unintelligible.

Otherwise we can certainly never imagine how so many instincts, and these so complex and perfect, could have issued suddenly "from the urn of hazard."

But Fabre will suppose nothing; he will only record the facts. Instead of wandering in the region of probabilities, he prefers to confine himself to the reality, and for the rest to reply simply that "we do not know."

This stern, positive, rigorous, independent, and observant mind, nourished upon geometry and the exact sciences, which has never been able to content itself with approximations and probabilities, could but distrust the seductions of hypotheses.

His robust common sense, which was always his protection against precipitate conclusions, too clearly comprehends the limits of science and the necessity of accumulating facts "upon the thorny path of observation and experiment" to indulge in generalization. He feels that life has secrets which our minds are powerless to probe, and that "human knowledge will be erased from the archives of the world before we know the last word concerning the smallest fly."

This is why he was regarded as "suspect" by the company of official scientists, to whom he was a dissenter, almost a traitor, especially at a moment when the theories of evolution, then in the first flush of their novelty, were everywhere the cause of a general elation.

CHAPTER 9.

No one as yet was capable of divining the man of the future in this modest thinker who would not accept the word of the masters interested, but in opposing the theory of transformation, far from being reactionary, Fabre revealed himself, at least in the domain of animal psychology, as an innovator, a true precursor.

Moreover, his observations, always so direct and personal, often revealed the contrary of what was asserted or foreseen by the magic formulae suggested by the mind.

To the ingenious mechanism invented by the transformists he preferred to oppose, not contrary argument, but the naked undeniable fact, the obvious testimony, the certain and irrefragable example. "Is it," he would ask them, "to repulse their enemies that certain caterpillars smear themselves with a corrosive product? But the larva of the Calosoma sycophanta, which feeds on the Processional caterpillar of the oak-tree, pays no heed to it, neither does the Dermestes, which feeds on the entrails of the Processional caterpillar of the pine-tree."

And consider mimicry. According to the theory of evolution, certain insects would utilize their resemblance to certain others in order to conceal themselves, and to introduce themselves into the dwellings of the latter as parasites living at their expense. Such would be the case with the

CHAPTER 9.   204

Volucella, a large fly whose costume, striped with brown and yellow bands, gives it a rude resemblance to the wasp. Obliged, if not for its own sake at least for that of its family, to force itself into the wasp's dwelling as a parasite, it deceitfully dresses itself, we are told, in the livery of its victim, thus affording the most curious and striking example of mimicry; and naturalists insufficiently informed would regard it as one of the greatest triumphs of evolution.

Now what does the Volucella do? It is true that it lays its eggs without being disturbed in the nest of the wasp. But, as the rigorous observer will tell you, it is a precious auxiliary and not an enemy of the community. Its grubs, far from disguising or concealing themselves, "come and go openly upon the combs, although every stranger is immediately massacred and thrown out." Moreover, "they watch the hygiene of the city by clearing the nest of its dead and ridding the larvae of the wasps of their excretory products." Plunging successively into each chamber of the dormitory the forepart of their bodies, "they provoke the emission of that fluid excrement of which the larvae, owing to their cloistration, contain an extreme reserve." In a word, the grubs of the Volucella "are the nurses of the larvae," performing the most intimate duties." (9/1.)

What an astonishing conclusion! What a disconcerting and unexpected reply to the "theories in vogue"!

Fabre, however, with his poetic temperament and ardent imagination, seemed admirably prepared to grasp all that vast network of relations by which all creatures are connected; but what proves the solidity of his imperishable work is that all theories, all doctrines, and all systems may resort to it in turn and profit by his proofs and arguments.

And he himself, although he boasts with so much reason of putting forward no pretensions, no theories, no systems, has he not even so yielded somewhat to the suggestions of the prevailing school of thought, and have not his verdicts against evolution often been the more excessive in that he has paid so notable a tribute to the evolutionary progress of creation?

In the first place, he is far from excluding the undeniable influence of environing causes; the immense role of those myriad external circumstances on which Lamarck so strongly insisted; but the work of these factors is, in his eyes, only accessory and wholly secondary in the economy of nature; and in any case it is far from explaining the definite direction and the transcendent harmony which characterize evolution, both in its totality and in its most infinitesimal details.

In one of his admirable little textbooks, intended to teach and to popularize science, he complacently enumerates the happy modifications effected by that "sublime magician," selection as understood by Darwin. He evokes the metamorphoses of the potato, which, on the mountains of Chili, is merely a wretched venomous tubercle, and those of the cabbage, which on the rocky face of oceanic precipices is nothing but a weed, "with a tall stem and scanty disordered leaves of a crude green, an acrid savour, and a rank smell"; he speaks of wheat, formerly a poor unknown grass; the primitive pear-tree "an ugly intractable thorny bush, with detestable bitter fruit"; the wild celery, which grows beside ponds, "green all over, hard, with a repulsive flavour, and which gradually becomes tenderer, sweeter, whiter," and "ceases to distil its poison." (9/2.)

With profound exactitude this great biologist has also perceived the degree to which size may be modified; may dwindle to dwarfness when a niggardly soil refuses to furnish beast and plant alike with a sufficient nourishment.

Without any communication with the other scientists who were occupied by the same questions, knowing nothing of the results which these experimenters had attained in the case of small mammiferous animals, and which prove that dwarfness has often no other cause than physiological poverty, he confirmed and expanded their ideas from an entomological point of view. (9/3.)

Scarcely ever, indeed, was he first inspired by the doings of others in this or that direction; he read scarcely anything, and nature was his sole teacher. He considered that the knowledge to be obtained from books is but so much vapour compared with the realities; he borrowed only from himself, and resorted directly to the facts as nature presented them. One has only to see his scanty library of odd volumes to be convinced how little he owes to others, whether writers or workers.

A true naturalist philosopher, this profound observer has also thrown a light upon certain singular anomalies which, in the insect world, seem to constitute an exception, at all events in our Europe, to the general rules. It is not only to the curiosity and for the amusement of entomologists that he proposes these curious anatomical problems, but also, and chiefly, to the Darwinian wisdom of the evolutionists.

Why, for example, is the Scarabaeus sacer born and why does it remain maimed all its life; that is to say, deprived of all the digits on the anterior limbs?

"If it is true that every change in the form of an appendage is only the sign of a habit, a special instinct, or a modification in the conditions of life, the theory of evolution should endeavour to account for this mutilation, for these creatures are, like all others, constructed on the same plan and provided with absolutely the same appendages."

The posterior limbs of the Geotrupes stercorarius, "perfectly developed in the adult, are atrophied in the larvae, reduced to mere specks."

The general history of the species, of its migrations and its changes, will doubtless one day throw light upon these strange infirmities, here temporary and there permanent, which may perhaps be explained by unforeseen encounters with undiscovered specimens, strayed perhaps into distant countries. (9/4.)

What invaluable documents for the entomologist and the historian of the evolution of the species are those multiple and fabulous metamorphoses of the Sitares and the Meloïdae which this indefatigable inquirer has revealed in all their astonishing phases!

One of the finest examples of scientific investigation is the pursuit, through a period of twenty-five years, with a sagacity which seems to border on divination, of this problem of HYPER-METAMORPHOSIS. The larvae of those coleoptera which we have seen introduced, with infernal cunning, into the cells of the Anthophora (See Chapter 8 above.), suffer no less than four moults before they become nymphs.

These merely external transformations, which involve only the envelope, and respect the internal structure, correspond each with a change of environment and of diet. Each time the organism adapts itself to its new mode of existence, "as perfectly as when it becomes adult"; and we see the insect, which was clear-sighted, become blind; it loses its feet, to recover them later; its slender body becomes ventripotent; hard, it grows soft; its mandibles, at first steely, become hollowed out spoonwise, each modification of conformation having its motive in a fresh modification of the conditions of the creature's life.

How explain this strange evolution of a fourfold larval existence, these successive appearances of organs, which become entirely unlike what they were, to serve functions each time different?

What is the reason, the intention, the high law which presides over these visible changes, these successive envelopments of creatures one within the other, these multiple transfigurations?

By what bygone adaptations has the Sitaris successively acquired these diverse extraordinary phases of life, indicating possibly for each corresponding age some ancient and remote heredity? (9/5.)

How many other arguments might evolution derive from his books, and what illustrations of the Darwinian philosophy has he unconsciously furnished! Does he not even allow the admission to escape him that "the spirit of cunning and deception is transmitted"? He sees in the persecutions of the Dytiscus, the "pirate of the ponds," the origin of the faculty which the Phryganea has of refashioning its shield when demanded of it. "To evade the assault of the brigand, the Phryganea must hastily abandon its mantle; it allows itself to sink to the bottom, and promptly removes itself; necessity is the mother of invention." (9/6.)

Returning to the lacunae which it so amazes Fabre to discover in our organization, even in the most perfect of us, are they fundamentally very real? These mysterious and unknown senses which he has so greatly contributed to elucidate in the case of the inferior species: why, he asks, have we not inherited them, if we are truly the final term and the supreme goal of creation?

But in cultivating our intuition, as Bergson invites us to do, would it be impossible to re-awaken, deep within us, these strange faculties, which perhaps are only slumbering? What of that species of indefinable memory which permits the red ant, the Bembex, the Cerceris, the Pompilus, the Chalicodoma and so many others to "find themselves," to orientate themselves with infallible certainty and incredible accuracy? Is it not to be found, according to travellers, in those men who have remained close to nature and accustomed from their remotest origins to listen to the silence of the great deserts?

Finally, the evolutionists, who "reconstruct the world in imagination," and who see in the relationship of neighbouring species a proof of descent or derivation, and a whole ideal series, will not fail to perceive throughout his work, in the elementary operations of the Eumenes and the Odynerus, cousins of the Cerceris, which sting their prey in places as yet ill determined, not indeed so many isolated attempts, but an incomplete process of invention, an attempt at procedures still in the fact of formation: in a word, the birth of that marvellous instinct which ends in the transcendent art of the Sphex and the Ammophila.

Although they have acquired such prodigious deftness, these master paralysers are not, in fact, always infallible. Occasionally the Sphex blunders and gropes, "operates clumsily"; the cricket revives, gets upon its feet, turns round and round, and tries to walk. But, inquires Fabre, do you say that having profited by a fortuitous act, which has turned out to be favourable to them, they have perfected themselves by contact with their elders, "thanks to the imitation of example," and that they have thus crystallized their experiences, which have been transmitted by heredity-- thereby fixed in the race? (9/7.)

How much we should prefer that it were so! How much more comprehensible and interesting their life would become!

But "when the hymenopteron breaks its cocoon, where are its masters! Its predecessors have long ago disappeared. How then can it receive education by example?"

You who "shape the world to your whim," you will reply: "Doubtless there are no longer masters to-day; but go back to the first ages of the globe, when the world in its newness, as Lucretius has so superbly said, as yet knew neither bitter cold nor excessive heat (9/8.); an eternal springtide bathed the earth, and the insects, not dying, as to-day, at the first touch of frost, two successive generations lived side by side, and the younger generation could profit at leisure by the lessons of example." (9/9.)

Let us return to Fabre's laboratory, to the covers of wire-gauze, and note what becomes, at the approach of winter, of the survivors of the vespine city.

In the mild and comfortable retreat where the wasps are kept under observation they die no less, despite their well-being and all the care expended on them, when once "the inexorable hour" has struck, and once the exact capital of life which seems to have been imparted to them ages ago is exhausted. With no apparent cause, we see death busy among them. "Suddenly the wasps begin to fall as though struck by lightning; for a few moments the abdomen quivers and the legs gesticulate, then finally remain inert, like a clockwork machine whose spring has run down to the last coil." (9/10.) This law is general; "the insect is born orphaned both of mother and father, excepting the social insect, and again excepting the dung-beetle, which dies full of days." (9/11.)

Moreover, Fabre is never weary of demonstrating that the insect, perfectly unconscious of the motive which makes it act, this thereby incapable of profiting by the lessons of experience and of innovation in its habits, beyond a very narrow circle. "No apprentices, no masters." In this world each obeys "the inner voice" on its own account; each sets itself to accomplish its task, not only without troubling as to what its neighbour is doing, but without thinking any further as to what it is doing itself; instance the Epeïra, turning its back on its work, yet "the latter proceeds of itself, so well is the mechanism devised"; and if by ill chance the spider acted otherwise it would probably fail.

Darwin knew barely the tenth part of the colossal work of Fabre. He had read firstly in the "Annals of Natural

Science" of the habits of the Cerceris and the fabulous history of the Meloidae. Finally he saw the first volume of the "Souvenirs" appear, and was interested in the highest degree by the beautiful study on the sense of location and direction in the Mason- bees.

This was already more than enough to excite his curiosity and to make him wonder whether all his philosophy would not stumble over this obstacle.

After having succeeded in explaining so luminously--and with what a lofty purview--the origin of species and the whole concatenation of animal forms, would it not be as though he halted midway in his task were the sanctuary of the origin of instinct to remain for ever inscrutable?

Fabre had not yet left Orange when Darwin engaged in a curious correspondence which lasted until the former had been nearly two years at Sérignan, and which showed how passionately interested the great theorist of evolution was in all the Frenchman's surprising observations.

It seems that on his side Fabre took a singular interest in the discussion on account of the absolute sincerity, the obvious desire to arrive at the truth, and also the ardent interest in his own studies, of which Darwin's letters were full. He conceived a veritable affection for Darwin, and commenced to learn English, the better to understand him and to reply more precisely; and a discussion on such a subject between these two great minds, who were, apparently, adversaries, but who had conceived an infinite respect for one another, promised to be prodigiously interesting.

Unhappily death was soon to put an end to it, and when the solitary of Down expired in 1882 the hermit of Sérignan saluted his great shade with real emotion. How many times have I heard him render homage to this illustrious memory!

But the furrow was traced; thenceforth Fabre never ceased to multiply his pin-pricks in "the vast and luminous balloon of transformism (evolution), in order to empty it and expose it in all its inanity." (9/12.) By no means the least original feature of his work is this passionate and incisive argument, in which, with a remarkable power of dialectic, and at times in a tone of lively banter, he endeavoured to remove "this comfortable pillow from those who have not the courage to inquire into its fundamental nature." He attacked these "adventurous syntheses, these superb and supposedly philosophic deductions," all the more eagerly because he himself had an unshakable faith in the absolute certainty of his own discoveries, and because he asserted the reality of things only after he had observed and re-observed them to satiety.

CHAPTER 9.

This is why he cared so little to engage in argument relating to his own works; he did not care for discussion; he was indifferent to the daily press; he avoided criticism and controversy, and never replied to the attacks which were made upon him; he rather took pains to surround himself with silence until the day when he felt that his researches were ripe and ready for publicity.

He wrote to his dear friend Devillario, shortly after Darwin's death:

"I have made a rule of never replying to the remarks, whether favourable or the reverse, which my writings may evoke. I go my own gait, indifferent whether the gallery applauds or hisses. To seek the truth is my only preoccupation. If some are dissatisfied with the result of my observations- -if their pet theories are damaged thereby--let them do the work themselves, to see whether the facts tell another story. My problem cannot be solved by polemics; patient study alone can throw a little light on the subject. (9/13.)

"I am profoundly indifferent to what the newspapers may say about me," he wrote to his brother seventeen years later; "it is enough for me if I am pretty well satisfied with my own work." (9/14.)

He read all the letters he received only in a superficial manner, neglecting to thank those who praised or congratulated him, and above all shrinking from all that idle correspondence in which life is wasted without aim or profit.

"I fume and swear when I have to cut into my morning in order to reply to so-and-so who sends me, in print or manuscript, his meed of praise; if I were not careful I should have no time left for far more important work."

His beloved Frédéric, "the best of his friends," was himself often treated no better, and to excuse his silence and the infrequency of his letters, Henri, even in the years spent at Carpentras and Ajaccio, could plead only the same reasons; his stupendous labours, his exhausting task, "which overwhelmed him, and was often too great, not for his courage, but for his time and his strength." (9/15.)

Nevertheless, while evading the question of origins, his far-sighted intellect was bound to "read from the facts" concerning the genesis of new species in process of evolution; and his observations throw a singular light on the quite recent theory of sudden mutations.

The nymph of the Onthophagus presents "a strange paraphernalia of horns and spurs which the organism has produced in a moment of ardour--a luxurious panoply which vanishes in the adult."

The nymph of the Oniticella also decks itself in "a temporary horn, which departs when it emerges."

And "as the dung-beetle is recent in the general chronology of creatures, as it takes rank among the last comers, as the geological strata are mute concerning it, it is possible that these horn-like processes, which always degenerate before they reach completion, may be not a reminiscence but a promise, a gradual elaboration of new organs, timid attempts which the centuries will harden to a complete armour, AND IF THIS WERE SO THE PRESENT WOULD TEACH US WHAT THE FUTURE IS TO BE." (9/16.)

Here is a specific transformation, a veritable creation; fortuitous, blind, and silent; one of those innumerable attempts which nature is always making, for the moment a mere matter of hazard, until some propitious circumstance fixes it in future incarnations.

Thus millions of indeterminate creatures are incessantly roughed out in the substance of that microcosm which is the initial cell; and it is here that Fabre sees the real secret of the law of evolution.

CHAPTER 9.

He refutes the great principle of Leibnitz, which was so brilliantly adopted by Darwin, that changes occur by degrees, by "fine shades," by slow variations, as the result of successive adaptations, and that there is no jumping-off place in nature. On the contrary, life often passes suddenly from one form to another, by abrupt and capricious leaps, by irregular and disorderly steps, and it is in the egg that Fabre sees the first lineaments of these mysterious and spontaneous variations.

Species are therefore born as a whole, each at the same time, AT THE SAME MOMENT, "bringing into being its new organism, with its individual properties and peculiarities, its indelible and innate faculties and tendencies, like "so many medals, each struck with a different die, which the gnawing tooth of time attacks only sooner or later to annihilate it."

However, Fabre affirms the continuity of progress; he believes in a better and more merciful future, a more complete humanity, ruled by more harmonious or less brutal laws.

With what profound intelligence and what generous enthusiasm he seeks to conjecture what this future might be, in his beautiful observations on the young of the Lycosa (9/17.), which can live for weeks and months in absolute abstinence, although we can perceive no reserve of nutriment!

We know no other sources of animal activity save the energy derived from food. Vegetables draw the materials of their nourishment from the soil and the air, and the sunlight is only an intermediary which enables the plant to fix its carbon. The animal species in turn borrow the elements indispensable to their existence from the vegetable world, or restore their flesh and blood with the flesh and blood of other animals.

Now the young Lycosae "are not inert on their mother's back; if they fall from the maternal chine they quickly pick themselves up and climb up one of her legs, and once back in place they have to preserve the equilibrium of the mass. In reality they know no such thing as complete repose. What then is the energetic aliment which enables the little Lycosae to struggle? Whence is the heat expended in action derived?"

Fabre sees no other source than "the sun."

"Every day, if the sky is clear, the Lycosa, loaded with her little ones, crawls to the edge of her well, and for long hours lies in the sun. There, on the maternal back, the young ones stretch themselves out, saturate themselves in the sunshine, charging themselves with motor reserves, steeping themselves in energy, directly converting into movement the calorific radiations coming from the sun, the centre of all life."

The Scorpion also is able to live for months without nourishment, restoring directly, in the form of movement, "the effluvia emanating from the sun or from other ambient energies--heat, electricity, light--which are the soul of the world."

Perhaps, among the innumerable worlds of space, there is somewhere, gravitating round a fixed star, a planet invisible to us where "the sunlight sates the hunger of the blind."

The gentle philosophy of the ingenious dreamer soothes itself with the vision, entertained by great and noble minds, of a humanity "whose teeth will no longer attack sensible life, nor even the pulp of fruits"; "when creatures will devour one another no longer, will no longer feed upon the dead; when they will be nourished by the sunlight, without conflict, without war, without labour; freed from all care, and assured against all needs!"

Thus, in the humblest creatures, he sees the most marvellous perspectives; the body of the lowest insect becomes suddenly a transcendent secret, lighting up the abyss of the human soul, or giving it a glimpse of the stars.

And although his work is in contradiction to the theories of the evolutionists, it ends with the same moral conclusion, namely, that all creation moves slowly and without intermission on its gradual ascent towards progress.

CHAPTER 10.

THE ANIMAL MIND.

The cunning anatomist has now successively laid bare all the springs of the animal intellect; he has shown how the various movements are mutually combined and engaged. But so far we have seen only one of the faces of the little mind of the animal; let us now consider the other aspect, the moral side, the region of feeling, the problem of which is confounded with the problem of instinct, and is doubtless fundamentally only another aspect of the same elemental power.

After the conflict the insect manifests its delight; it seems sometimes to exult in its triumph; "beside the caterpillar which it has just stabbed with its sting, and which lies writhing on the ground," the Ammophila "stamps, gesticulates, beats her wings," capers about, sounding victory in an intoxication of delight.

The sense of property exists in a high degree among the Mason-bees; with them right comes before might, and "the intruder is always finally dislodged." (10/1.)

But can we find in the insect anything analogous to what we term devotion, attachment, affectionate feeling? There are facts which lead us to believe we may.

Let us go once more into Fabre's garden and admire the Thomisus: absorbed in her maternal function, the little spider lying flat on her nest can strive no longer and is wasting away, but persists in living, mere ruin that she is, in order to open the door to her family with one last bite. Feeling under the silken roof her offspring stamping with impatience, but knowing that they have not strength to liberate themselves, she perforates the capsule, making a sort of practicable skylight. This duty accomplished, she quietly surrenders to death, still grappled to her nest.

The Psyche, dominated by a kind of unconscious necessity, protects her nursery by means of her body, anchors herself upon the threshold, and perishes there, devoted to her family even in death.

However, Fabre will show us with infallible logic that all these instances of foresight and maternal tenderness have, as a rule, no other motive than pleasure and the blind impulse which urges the insect to follow only the fatal path of its instincts.

In many species the material fact of maternity is reduced to its simplest expression.

CHAPTER 10.

The Pieris limits herself to depositing her eggs on the leaves of the cabbage, "on which the young must themselves find food and shelter."

"From the height of the topmost clusters of the centaury the Clythris negligently lets her eggs fall to the ground, one by one, here or there at hazard; without the least care as to their installation.

"The eggs of the Locustidae are implanted in the earth like seeds and germinate like grain."

But stop before the Lycosa, that magnificent type of maternal love which Fabre has already depicted. "She broods over her eggs with anxious affection. With the hinder claws resting on the margin of the well she holds herself supported above the opening of the white sac, which is swollen with eggs. For several long weeks she exposes it to the sun during half the day. Gently she turns it about in order to present every side to the vivifying light. The bird, in order to hatch her eggs, covers them with the down of her breast, and presses them against that living calorifer, her heart. The Lycosa turns hers about beneath the fires of heaven; she gives them the sun for incubator." (10.2.) Could abnegation be more perfect? What greater proof could there be of renunciation and self-oblivion?

But appearances are vain. Substitute for the beloved sac some other object, and the spider "will turn about, with the same love, as though it were her sac of eggs, a piece of cork, a pincushion, or a ball of paper," just as the hen, another victim of this sublime deception, will give all her heart to hatching the china nest-eggs which have been placed beneath her, and for weeks will forget to feed.

The young brood hatches, and the spider goes a-hunting, carrying her little ones on her back; she protects them in case of danger, but is incapable of recognizing them or of distinguishing them from the young of others. The Copris and the Scorpion are no less blind, "and their maternal tenderness barely exceeds that of the plant, which, a stranger to any sense of affection or morality, none the less exercises the most exquisite care in respect of its seeds."

Moreover, the impulse to work is only a kind of unconscious pleasure. When the Pelopaeus "has stored her lair with game," when the Cerceris has sealed the crypt to which she has confided the future of her race, neither one nor the other can foresee "the future offspring which their faceted eyes will never behold, and the very object of their labours is to them occult."

With them, as with all, life can only be a perpetual illusion.

Yet the marvellous edifice of the "Souvenirs entomologiques" is consummated by the astonishing history of the Minotaur, whose habits surpass in ideal beauty all that could be imagined.

At the bottom of a burrow, in a deeply sunken vault, two dung-beetles are at work, the Minotaurs, who, once united, recognize one another, and can find one another again if separated, but do not voluntarily separate, realizing "the moral beauty of the double life" and "the touching concept of the family, the sacred group par excellence." The male buries himself with his companion, remains faithful to her, comes to her assistance, and "stores up treasure for the future. Never discouraged by the heavy labour of climbing, leaving to the mother only the more moderate labour, keeping the severest for himself, the heavy task of transport in a narrow tunnel, very deep and almost vertical, he goes foraging, forgetful of himself, heedless of the intoxicating delights of spring, though it would be so good to see something of the country, to feast with his brothers, and to pester the neighbours; but no! he collects the food which is to nourish his children, and then, when all is ready for the new-comers, when their living is assured, having spent himself without counting the cost, exhausted by his efforts, and feeling himself failing, he leaves his home and goes away to die, that he may not pollute the dwelling with a corpse."

The mother, on her side, allows nothing to divert her from her household, and only returns to the surface when accompanied by her young, who disperse at will. Then, having nothing more to do, the devoted creature perishes in turn. (10/3.)

Compared with the Scarabaeus, which contents itself with idle wandering, or even with the meritorious Sisyphus, does it not seem that the Minotaur moves on an infinitely higher plane?

What nobler could be found among ourselves? What father ever better comprehended his duties and obligations toward his family? What morality could be more irreproachable; what fairer example could be meditated?

"Is not life everywhere the same, in the body of the dung-beetle as in that of man? If we examine it in the insect, do we not examine it in ourselves?"

Whence does the Minotaur derive these particular graces? How has it risen to so high a level on the wings of pure instinct? How could we explain the rarity of so sublime an example, did we not know, to satiety, that "nature everywhere is but an enigmatic poem, as who should say a veiled and misty picture, shining with an infinite variety of deceptive lights in order to evoke our conjectures"? (10/4.)

Nevertheless, it is a fact that the majority have no other rule of conduct than to follow the trend of their instincts, and to obey "their unbridled desires." No one better than Fabre has expounded the blind operation of these little natural forces, the brutality of their manners, their cannibalism, and what we might call their amorality, were it possible to employ our human formulae outside our own human world.

With the gardener-beetles, if one is crippled, none of the same race halts or lingers; none attempts to come to his aid. Sometimes the passers-by hasten to the invalid to devour him."

In the republic of the wasps "the grubs recognized as incurable are pitilessly torn from their place and dragged out of the nest. Woe to the sick! they are helpless and at once expelled."

When the winter comes all the larvae are massacred, and the whole vespine city ends in a horrible tragedy.

But life is a whole, and all conduct is good whose actions realize an object and are adapted to an end. If there is a "spirit" of the hive, the insect also has its morality and the wasp's nest its "law," and the conduct of its inmates, horrible though it may seem to Fabre, is doubtless only a submission to certain exigencies of that universal law which makes nature a "savage foster-mother who knows nothing of pity."

These cruelties particularly show us that one of the functions of the insect in nature is to preside over the disappearance and also the ultimate metamorphoses of the least "remnants of life."

Each has its providential hygienic function.

The Necrophori, "the first of the tiny scavengers of the fields," bury corpses in order to establish their progeny in them; in the space of a few hours an enormous body, a mole, a water-rat, or an adder, will completely disappear, buried under the earth.

The Onthophagi purify the soil, "dividing all filth into tiny crumbs, ridding the earth of its defilements."

A very small beetle, the Trox, has the imprescriptible mission of purging the earth of the rabbits' fur rejected by the fox. (10/5.)

Here structure explains the function.

CHAPTER 10.

The intestine of the grub of the rose-beetle "is a veritable triturating mill, which transforms vegetable matter into mould; in a month it will digest a volume of matter equal to several thousand times the initial volume of the grub."

The intestine of the Scarabaei is prolonged to a prodigious length in order to "drain the excrement to the last atom in its manifold circuits. The sheep has finely divided the vegetable matter; the grub, that incomparable triturator, reduces it to the finest possible consistency; not a morsel is left in which the magnifying glass can reveal a fibre."

To fulfil its hygienic mission the insect arrives in due season, and multiplies its legions; "there are twenty thousand eggs in the flanks of the house fly; immediately they are hatched these twenty thousand maggots set to work, so that Linnaeus has said that three flies would suffice to devour the body of a horse or a lion."

Feeding only upon wheat, a single weevil, the Calendar beetle, produces ten thousand eggs, whence issue as many larvae, each of them devouring its grain.

In all species the number of births is at first exaggerated, for all, the obscure, the nameless, the most destructive, our pests as well as our most precious helpers, have their utility and their part to play in the general scheme of life, a raison d'être in the eternal renewal of things, which is without reference to the vexatious or beneficent quality of their behaviour to us.

Each has its rank assigned, each has its task, to one the flower, to another the roots, to a third the leaves; the vine has its caterpillars, its beetles, its butterflies; the clover, its moths and mites. (10/6.)

Man sees himself forced to submit to them, and spends himself in vain efforts to carry on an often useless campaign. Nothing seems to affect them, neither drought, nor rain, nor even the severest cold; and the eggs and larvae, organizations apparently delicate in the extreme, are often more tenacious of life than the adults. Fabre has proved this: let the temperature suddenly fall twenty degrees: the eggs of Geotrupes and the larvae of the cockchafer or the rose-beetle endure such vicissitudes of temperature with impunity; contracted and stiffened into little masses of ice, but not destroyed, they revive in spring no less than the eel fry, the rotifers, or the tardigrades. One can scarcely believe that life still persists in a state of suspense only in these little frozen creatures, whose organization is already so complicated.

Then, of a sudden, the ravagers disappear; more often than not none knows how or why; deliverance is at hand.

What indeed would become of the world were nothing to moderate such fecundity?

Again, each species has its trials which appear in time to moderate its surplusage, and Fabre expounds for us, with a stern philosophy, the terrible devices by which this repression is effected.

Each has its appointed enemy, which lives upon it or its offspring, and which in turn becomes the prey of some smaller creature. The gentle itself, "the king of the dead," has its parasites. While it swims in the deliquescence of putrefying flesh a minute Chalcidian perforates its skin with an imperceptible wound, and introduces its terrible eggs, whence in the future will issue larvae which to-morrow will devour the devourers of to-day.

None exists save to the detriment of others. Everywhere, even in the smallest, we find "an atrocious activity, a cunning brigandage," a savage extermination, which dominates a vast unconscious world of which the final result is the restoration of equilibrium. (10/7.) It is only on these antagonisms, on the enemies of our enemies, that we can found any hope of seeing this or that pest disappear. A small Hymenopteron, almost invisible, the Microgaster glomeratus, is entrusted with the destruction of the cabbage caterpillar; the cochineal wages war to the death upon the green- fly; the Ammophila is the predestined murderer of the harvest Noctuela, whose misdeeds in a beetroot country often amount to a disaster. The Odynerus has for its instinctive mission to arrest the excessive multiplication of a lucerne weevil, no less than twenty-four of whose grubs are necessary to rear the offspring of the brigand, and nearly sixty gadflies are sacrificed to the growth of a single Bembex.

Everywhere craft is organized to triumph over force. Around each nest the parasites lie in wait, "atrocious assassins of the child in the cradle, watching at the doors for the favourable occasion to establish their family at the expense of others. The enemy penetrates the most inaccessible fortress; each has its tactics of war, devised with a terrible art. Of the nest and the cocoon of the victim the intruder makes its own nest, its own cocoon, and in the following year, instead of the master of the house, he will emerge from underground as the usurping bandit, the devourer of the inhabitant."

While the cicada is absorbed in laying her eggs an insignificant fly labours to destroy them. How express the calm audacity of this pigmy, following closely after the colossus, step by step; several at once almost under the talons of the giant, which could crush them merely by treading on them? But the cicada respects them, or they would long ago have disappeared." (10/8.)

Fabre thus agrees with Pasteur, who in the world of the infinitely little shows us the same antagonisms, the same vital competition, the same eternal movement of flux and reflux, the same whirlpool of life, which is extinguished only to reappear: tending always towards an equilibrium which is incessantly destroyed. And it is thanks to this balancing that the integral of life remains everywhere and always almost identical with itself.

CHAPTER 11.

HARMONIES AND DISCORDS.

Such indeed is the economy of nature that secret relations and astonishing concordances exist throughout the whole vast weft of things. There are no loose ends; everything is consequent and ordered. Hidden harmonies meet and mingle.

Among the terebinth lice, "when the population is mature, the gall is ripe also, so fully do the calendars of the shrub and the animal coincide"; and the mortal enemy of the Halictus, the sinister midge of the springtime, is hatched at the very moment when the bee begins to wander in search of a location for its burrows.

The fantastic history of the larvae of the Anthrax furnishes us with one of the most suggestive examples of these strange coincidences. (10/9.)

The Anthrax is a black fly, which sows its eggs on the surface of the nests of the Mason-bee, whose larvae are at the moment reposing in their silken cocoons.

"The grub of the Anthrax emerges and comes to life under the touch of the sunlight. Its cradle is the rugged surface of

CHAPTER 11.

the cell; it is welcomed into the world by a literally stony harshness...Obstinately it probes the chinks and pores of the nest; glides over it, crawls forward, returns, and recommences. The radicle of the germinating seed is not more persevering, not more determined to descend into the cool damp earth. What inspiration impels it? What compass guides it? What does the root know of the fertility of the soil?...The nurseling, the seed of the Anthrax, is barely visible, almost escaping the gaze of the magnifying glass; a mere atom compared to the monstrous foster-mother which it will drain to the very skin. Its mouth is a sucker, with neither fangs nor jaws, incapable of producing the smallest wound; it sucks in place of eating, and its attack is a kiss." It practises, in short, a most astonishing art, "another variation of the marvellous art of feeding on the victim without killing it until the end of the meal, in order always to have a store of fresh meat. During the fourteen days through which the nourishment of the Anthrax continues, the aspect of the larva remains that of living flesh; until all its substance has been literally transferred, by a kind of transpiration, to the body of the nurseling, and the victim, slowly exhausted, drained to the last drop, while retaining to the end just enough life to prove refractory to decomposition, is reduced to the mere skin, which, being insufflated, puffs itself out and resumes the precise form of the larva, there being nowhere a point of escape for the compressed air."

CHAPTER 11. 238

Now the grub of the Anthrax "appears precisely at the exact moment when the larva of the Chalicodoma is attacked by that lethargy which precedes metamorphosis, and which renders it insensible, and during which the substance of the grub about to be transfigured into a bee commences to break down and resolve itself into a liquid pulp, for the processes of life always liquefy the grub before achieving the perfect insect." (11/2.)

Here again the time-tables coincide.

But it is perhaps in the celebrated Odyssey of the grub of the Sitaris that Fabre most urgently claims our admiration for the marvellous and incomprehensible wisdom of the Unconscious!

Let us recapitulate the unheard-of series of events, the inextricable complication of circumstances, which are required to condition the lowly life of a Sitaris.

In the first place, this microscopic creature must be provided with talons, or how could it adhere to the fleece of the Anthophora, on which it must live as parasite for a certain length of time?

Then again, it must transfer itself from the male to the female bee in the course of its travels abroad, or its destiny would be cut short.

Again, it must not miss the opportunity of embarking itself upon the egg just at the propitious moment.

Then the volume of this egg must be so calculated as to represent an allowance of food exactly proportioned to the duration of the first phase of its metamorphosis. Moreover, the quantity of honey accumulated by the bee must suffice for the whole of the remaining cycle of its larval existence.

Let a single link of the chain be broken, and the entire species of the Sitaris is no longer possible.

If every species has its law; if the Geotrupes remain faithful to filth, although experience shows that they can accommodate themselves equally well to the putrefaction of decayed leaves; if the predatory species--the Cerceris, the Sphex, the Ammophila--resort only to one species of quarry to nourish their larvae, although these same larvae accept all indifferently, it is on account of those superior economic laws and secret alliances the profound reasons for which as a rule escape us or are beyond the scope of our theories.

For all things are produced and interlocked by the eternal necessity; link engages in link, and life is only a plexus of solitary forces allied among themselves by their very nature, the condition of which is harmony. And the whole system of living creatures appears to us, through the work of the great naturalist, as an immense organism, a sort of vast physiological apparatus, of which all the parts are mutually interdependent, and as narrowly controlled as all the cells of the human body.

Fabre goes on to present us with other facts, which at a first glance appear highly immoral; I am referring to certain phases of sexual love among the lower animals, and his ghoulish revelations concerning the horrible bridals of the Arachnoids, the Millepoda, and the Locustidae.

The Decticus surrenders only to a single exploit of love; a victim of its "strange genesics"; utterly exhausted by the first embrace, empty, drained, extenuated, motionless in all its members, utterly worn out, it quickly succumbs, a mere broken simulacrum, like the miserable lover of a monstrous succubus who "loves him enough to devour him." (11/3.)

The female scorpion devours the male; "all is gone but the tail!"

The female Spider delights in the flesh of her lover.

The cricket also devours a small portion of her "debonair" admirer.

The Ephippigera "excavates the stomach of her companion and eats him."

But the horror of these nuptial tragedies is surpassed by the insatiable lust, the monstrous conjunction, the bestial delights of the Mantis, that "ferocious spectre, never wearied of embraces, munching the brains of its spouse at the very moment of surrendering her flanks to him." (11/4.)

Whence these strange discords, these frightful appetites?

Fabre refers us to the remotest ages, to the depths of the geological night, and does not hesitate to regard these cruelties as "remnants of atavism," the lingering furies of an ancient strain, and he ventures a profound and plausible explanation.

The Locusts, the Crickets, and the Scolopendrae are the last representatives of a very ancient world, of an extinct fauna, of an early creation, whose perverse and unbridled instincts were given free vent, when creation was as yet but dimly outlined, "still making the earliest essays of its organizing forces"; when the primitive Orthoptera, "the obscure forebears of those of to-day, were "sowing the wild oats of a frantic rut, "in the colossal forests of the secondary period; by the borders of the vast lakes, full of crocodiles, and antediluvian marshes, which in Provence were shaded by palms, and strange ferns, and giant Lycopodia, never as yet enlivened by the song of a bird.

These monstrosities, in which life was making its essays, were subject to singular physical necessities. The female reigned alone; the male did not as yet exist, or was tolerated only for the sake of his indispensable assistance. But he served also another and less obvious end; his substance, or at least some portion of his substance, was an almost necessary ingredient in the act of generation, something in the nature of a necessary excitant of the ovaries, "a horrible titbit," which completed and consummated the great task of fecundation. Such, in Fabre's eyes, was the imperious physiological reason of these rude laws. This is why the love of the males is almost equivalent to their suicide; the Gardener-beetle, attacked by the female, attempts to flee, but does not defend himself; "it is as though an invincible repugnance prevents him from repulsing or from eating the eater." In the same way the male scorpion "allows himself to be devoured by his companion without ever attempting to employ his sting," and the lover of the Mantis "allows himself to be nibbled to pieces without any revolt on his part."

CHAPTER 11.

A strange morality, but not more strange than the organic peculiarities which are its foundation; a strange world, but perhaps some distant sun may light others like it.

These terrible creatures are a source of dismay to Fabre. If all things proceed from an underlying Reason, if the divine harmony of things testifies everywhere to a sovereign Logic, how shall the proofs of its excellence and its sovereign wisdom be found in such things as these?

Far from attributing to the order of the universe a supposed perfection, far from considering nature as the most immediate expression of the Good and the Beautiful, in the words of Tolstoy (11/5.), he sees in it only a rough sketch which a hidden God, hidden, but close at hand, and living eternally present in the heart of His creatures, is seeking to test and to shape.

Living always with his eyes upon some secret of the marvels of God, whom he sees in every bush, in every tree, "although He is veiled from our imperfect senses" (11/6.), the vilest insect reveals to him, in the least of its actions, a fragment of this universal Intelligence.

What marvels indeed when seen from above! But consider the Reverse--what antinomies, what flagrant contradictions! What poor and sordid means! And Fabre is astonished, in spite of all his candid faith, that the fatality of the belly should have entered into the Divine plan, and the necessity of all those atrocious acts in which the Unconscious delights. Could not God ensure the preservation of life by less violent means? Why these subterranean dramas, these slow assassinations? Why has Evil, THE POISON OF THE GOOD (11/7.), crept in everywhere, even to the origin of life, like an eternal Parasite?

Within this fatal circle, in which the devourer and the devoured, the exploiter and the exploited, lead an eternal dance, can we not perceive a ray of light?

For what is it that we see?

The victims are not merely the predestined victims of their persecutors. They seek neither to struggle nor to escape nor to evade the inevitable; one might say that by a kind of renunciation they offer themselves up whole as a sacrifice!

What irresistible destiny impels the bee to meet half-way the Philanthus, its terrible enemy! The Tarantula, which could so easily withstand the Pompilus, when the latter rashly carries war into its lair, does not disturb itself, and never dreams of using its poisoned fangs. Not less absolute is the submission of the grasshopper before the

Mantis, which itself has its tyrant, the Tachytes.

Similarly those which have reason to fear for their offspring, if not for themselves, do nothing to evade the enemy which watches for them; the Megachile, although it could easily destroy it, is indifferent to the presence of a miserable midge, "the bandit who is always there, meditating its crime"; the Bembex, confronted with the Tachinarius, cannot control its terror, but nevertheless resigns itself, while squeaking with fright.

If each creature is what it is only because it is a necessary part of the plan of the supreme Artisan who has constructed the universe, why have some the right of life and death and others the terrible duty of immolation?

Do not both obey, not the gloomy law of carnage, but a kind of sovereign and exquisite sacrifice, some sort of unconscious idea of submission to a superior and collective interest?

This hypothesis, which was one day suggested to Fabre by a friend of great intellectual culture (11/8.), charmed and interested him keenly. I noticed that he was more than usually attentive, and he seemed to me to be suddenly reassured and appeased. For him it was as though a faint ray of light had suddenly fallen among these impenetrable and distressing problems.

It seemed to him that by setting before our eyes the spectacle of so many woes, universally distributed, and doubtless necessary, woes which do not spare even the humblest of creatures, the Sovereign Intelligence intends to exhort us to examine ourselves truly and to dispose us to greater love and pity and resignation.

All his work is highly and essentially religious; and while he has given us a taste for nature, he has not also endeavoured to give us, according to the expression of Bossuet "the taste for God," or at least a sense of the divine? In opposing the doctrine of evolution, which reduces the animal world to the mere virtualities of the cell; in revealing to us all these marvels which seem destined always to escape human comprehension; finally, by referring us more necessarily than ever to the unfathomable problem of our origins, Fabre has reopened the door of mystery, the door of the divine Unknown, in which the religion of men must always renew itself. We should belittle his thought, we should dwarf the man himself, were we to seek to confine to any particular thesis his spiritualistic conception of the universe.

Fabre recognizes and adores in nature only the great eternal Power, whose imprint is everywhere revealed by the phenomena of matter.

For this reason he has all his life remained free from all superstition and has been completely indifferent to dogmas and miracles, which to his mind imply not only a profound ignorance of science, but also a gross and complete miscomprehension of the divine Intelligence. He kneels upon the ground or among the grasses only the more closely to adore that force, the source of all order, the intuitive knowledge of which, innate in all creatures, even in the tiny immovable minds of animals, is merely a magnificent and gratuitous gift. The office in which he eagerly communicates is that glorious and formidable Mass in which the ragged sower, "noble in his tatters, a pontiff in shabby small-clothes, solemn as a God, blesses the soil, more majestic than the bishop in his glory at Easter-tide." (11/9.) It is there that he finds his "Ideal," in the incense of the perfumes "which are softly exhaled from the shapely flowers, from their censers of gold," in the heart of all creatures, "chaffinch and siskin, skylark and goldfinch, tiny choristers" piping and trilling, "elaborating their motets" to the glory of Him who gave them voice and wings on the fifth day of Genesis. He fraternizes with all, with his dogs and his cats, his tame tortoise, and even the "slimy and swollen frog"; the "Philosopher" of the Harmas, whose murky eyes he loves to interrogate as he paces his garden "by the light of the stars"; persuaded that all are accomplishing a useful work, and that all creatures, from the humblest insect which has only nibbled a leaf, or displaced a few grains of sand, to man himself, are anointed with the same chrism of immortality.

And as he has always set the pleasures of study before all others, he can imagine no greater recompense after death than to obtain from heaven permission still to continue in their midst, during eternity, his life of labour and effort.

CHAPTER 12.

THE INTERPRETATION OF NATURE.

We have noted the essential features of his precise and unfailing vision and the value of the documents which record the work of Fabre, but the writer merits no less attention than the observer and the philosopher.

In the domain of things positive, it is not always sufficient to gather the facts, to record them, and to codify in bare formulae the results of inquiry. Doubtless every essential discovery is able to stand by itself; in what would an inventor profit, for example, by raising himself to the level of the artist? "For the theorem lucidity suffices; truth issues naked from the bottom of a well."

But the manner of speaking, describing, and depicting is none the less an integral part of the truth when it is a matter of expounding and transmitting the latter. To express it feebly is often to compromise it, to diminish it; and even to betray it. There are terms which say better than others what has to be said. "Words have their physiognomy; if there are lifeless words, there are also picturesque and richly-coloured words, comparable to the brush strokes which scatter flecks of light on the grey background of the picture." There are particular terms of expression, felicities which present things in a better light, and the writer must search in his memory, his imagination, and his heart, for the fitting accent; for the flexibility of language and the wealth of words which are needful if he would fully succeed in the portrayal of living creatures; if he would tender the living truth, reproduce in all its light and shade the spectacle of the world, arouse the imagination, and faithfully interpret the mysterious spirit which impregnates matter and is reflected in thought.

The artist then comes forward to co-ordinate all these scattered fragments, to assemble them, to breathe vitality into them, to restore these inert truths to life.

But what a strange manner of working was Fabre's; what a curious method of composition! However full of ideas his mind might be, he was incapable of expressing them if he remained in one place and assumed the ordinary preliminary attitude of a man preparing to write. Seated and motionless, his limbs at rest, pen in hand, with a blank page before him, it seemed to him that all his faculties became of a sudden paralysed. He must first move about; activity helped him to pursue his ideas; it was in action that he recovered his ardour and uncovered the sources of inspiration. Just as he never observed without enthusiasm, so he found it impossible to write without exaltation, and it was precisely because he so ardently loved the truth that

CHAPTER 12.

he felt himself compelled to show it in all its beauty.

Moving like a circus-horse about the great table of his laboratory, he would begin to tramp indefatigably round and round, so that his steps have worn in the tiles of the floor an ineffaceable record of the concentric track in which they moved incessantly for thirty years.

His mind would grow clear and active as he walked, smoking his pipe and "using his marrow-bones." (12/1.) He was already at work; he was "hammering" his future chapters in his brain; for the idea would be all the more precise as the form was more finished and more irreproachable, more closely identified with the thought; he would wait until the word quivered, palpitated, and lived; until the transcription was no longer an illusion, a phantom, a vision devoid of reality, but a faithful echo, a sincere translation, a finished interpretation, reflecting entire the fundamental essence of the thing; in a word, a work of art, a parallel to nature.

Then only would he sit before the little walnut-wood table "spotted with ink and scarred with knife-cuts, just big enough to hold the inkstand, a halfpenny bottle, and his open notebook": that same little table at which, in other days, by force of meditation, he achieved his first degrees.

CHAPTER 12.

Then he would begin to write, "his pen dipped not in ink only" but in his heart's blood (12/2.); first of all in ordinary ruled notebooks bound in black cloth, in which he noted, day by day, hour by hour, the observations of every moment, the results of his experiments, together with his thoughts and reflections. Little by little those documents would come together which elucidated and completed one another, and at last the book was written. These notebooks, these copious records, are remarkable for the regularity of the writing and the often impeccable finish of the first draught. Although here and there the same data are transcribed several times in succession, and each time struck through with a vigorous stroke of the pen, there are whole pages, and many pages together, without a single erasure. The handwriting, excessively small--one might think it had been traced by the feet of a fly--becomes in later years so minute that one almost needs a magnifying glass to decipher it.

These notebooks are not the final manuscript. The entomologist would write a new and more perfect copy on loose sheets of paper, making one draught after another, patiently fashioning his style and polishing his work, although many passages were included without revision as they were written in the first instance.

CHAPTER 12. 253

The greatest magician of modern letters, versed in all the artifices of the French language, speaking one day of Fabre and his writings, made in my hearing the assertion that he was not, properly speaking, an artist. He might well be a great naturalist, a veteran of science, an observer of genius, but he was by no means and would never be a writer according to the canons of the craft.

But how many others, like him, in their time regarded as "pitiable in respect of their language," charm us to-day, simply because they were gifted with imagination and the power of giving life to their work! (12/3.)

To tell the truth, Fabre is absolutely careless of all literary procedure, and solely preoccupied with bringing his style into harmony with his thoughts; he is not in the least a manufacturer of literary phrases. There is no trace of artistic writing in his books, and it is only his manner of feeling and of expressing himself that makes him so dear to us.

What touches us in him is the accent, the simplicity, the measure, the good sense, and the perfect equilibrium of each of these pages: simple, often commonplace, even incorrect or trivial, but so alive, so human, that the blood seems to flow in them. It is the lover in Fabre that draws us to him; nothing quite like his work has been seen since the days of Jean de La Fontaine.

He has liberated science; he laughs at the specialists who take refuge behind their "barbarian terminologies," at the "jargon" of those "who see the world only through the wrong end of the glass"; at the exaggerated importance which they attribute to insignificant details, the narrowness of classifications, and the chaos of systems; all that incoherent, remote, and inaccessible science, which he, on the contrary, strives to render pleasant and attractive.

This is why the great scientist has endeavoured to speak like other people, preferring, to the harsh consonants of technical phrases which sound "like insults" or have the air of "a magical invocation, which make certain scientific works read like so much gibberish," the "naive and picturesque appellation, the familiar, trivial name, the popular, living term which directly interprets the exact signification of the habits of an insect, or informs us fully of its dominant characteristic, or which, at least, leaves nothing to conjecture."

He considers it useless and even inconvenient to abandon many charming expressions, appropriate and significant as they are, which may be borrowed from the good old French tongue; and in this he resembles the immortal de Jussieu, who in his botanical classifications was careful not to discard the old popular denominations which Theophrastus, Virgil, and Linnaeus had thought fit to bestow upon plant and tree.

It is for the same reasons that he loves the Provençal tongue; that beautiful idiom, that superb language, rich in music, in sonorous words, so suggestive and so full of colour, many of whose terms, saying precisely what they intend to say, have no equivalent in French. He has learned the language, and reads it: in particular Roumanille, whose easy, familiar style pleases him better than the grandiloquence of Mistral, although he delights also in Calendal, whose lyrical powers fill him with enthusiasm. >From this ancient tongue, which was early as familiar to him as the French, he borrowed certain mannerisms, certain tricks of style, certain neologisms, and also, to some extent, his simplicity of manner and the cadence of his prose.

It was not without difficulty that he attained this mastery. Measure the gulf between his first volumes and his last; in the first the style is slightly nerveless and indefinite: it was only as he gradually advanced in his career that he acquired what may be called his final manner, or achieved, in his narratives, a perfect literary style. The most substantially constructed, the most happily expressed of his pages were written principally in his extreme old age.

Not only is there no sign of failing in these, but in his latest "Souvenirs" the perfection of form is perhaps even more remarkable than the wealth of matter.

How vitally his scrupulous records impress the mind's eye; how firmly they establish themselves in the memory!

Even if one has never seen the Pelopaeus, one readily conceives an impression of "her wasp-like costume, and curving abdomen, suspended at the end of a long thread." What exactitude in this snapshot, taken at the moment when the insect is occupied in scooping out of the mire the lump of mud intended for the construction of her nest: "like a skilled housekeeper, with her clothing carefully tucked up that it may not be soiled, the wings vibrating, the limbs rigidly straightened, the black abdomen well raised on the end of its yellow stalk, she rakes the mud with the points of her mandibles, skimming the shining surface." (12/4.)

He draws, in passing, this charming sketch of the gadfly, the pest of horses, which nourishes itself with their blood:

"Gadflies of several species used to take refuge under the silken dome of my umbrella, and there they would quietly rest, one here, one there, on the tightly stretched fabric; I rarely lacked their company when the heat was overpowering. To while away the hours of waiting, I used to love to watch their great golden eyes, which would shine like carbuncles on the vaulted ceiling of my shelter; I used to love to watch them slowly change their stations, when the excessive heat of some point of the ceiling would force them to move a little." (12/5.)

We follow all the manoeuvres of the Balaninus, the acorn-weevil, "burying her drill" which "operates by means of little bites." The narrator calls our attention to the slightest episodes, even to those accidents which sometimes surprise the worker in the course of her labours; when, with the rostrum buried deep in the acorn, her feet suddenly lose their hold. Then the unhappy creature, unable to free herself, finds herself suspended in the air, at right angles to her proboscis, far from any foothold or point of vantage, at the extremity of her disproportionately long pike, that "fatal stake." (12/6.)

As for the poplar-weevil, we can almost see it moving "in the subtlest equilibrium, clinging with its hooked talons to the slippery surface of the leaf"; we watch all the details of its methods and the progress of its labours. We see the flexed leaf assume the vertical under the awl-stroke which the insect applies to the pedicle, "when, partially deprived of sap, the leaf becomes more flexible, more malleable; it is in a sense partly paralysed, only half alive." Then we follow the rolling process; "the imperturbable deliberation of the worker as it rolls its cigar, which finally hangs perpendicularly at the end of the bent and wounded stem." (12/7.)

Fabre, like a true artist, finds all sorts of expressions to describe the tiny, fragile eggs of his insects; little shining pearls, delicious coffers of nickel or amber, miniature pots of translucid alabaster, "which we might think were stolen from the cupboard of a fairy."

He opens the enchanted alcoves wherein the puny grubs lie slumbering, "fat, rounded puppets"; the tender larvae which "gape and swing their heads to and fro" when the mother returns to the nest with her toothsome mouthful or her crop swollen with honey.

What compassion, what tenderness, what sensitiveness in the affecting picture of the mother Halictus, abandoned, deprived of her offspring, bewildered and lost, when the terrible spring fly has destroyed her house: bald, emaciated, shabby, careworn, already dogged by the small grey lizard! (12/8.)

The tragedy of the wasps' nest at the approach of the first chills of winter is the final fragment of an epic. At first there is a sort of uneasiness, "a species of indifference and anxiety which broods over the city"; already it has a presentiment of coming misfortune, of an approaching catastrophe. Presently a wild excitement ensues; the foster- mothers, "frightened, fierce, and restless," as though suddenly attacked by an incomprehensible insanity, conceive an aversion for the young; "the neuters extirpate the larvae and drag them out of the nest," and the drama of destruction draws to a close with "the final catastrophe; the infirm and the dying are dismembered, eviscerated, dissected in a heap in the catacombs by maggots, woodlice, and centipedes." Finally the moth comes upon the scene, its larvae "attacking the dwelling itself; gnawing and destroying the joists and rafters, until all is reduced to a few pinches of dust and shreds of grey paper." (12/9.)

What picturesque expressions he employs to depict, by means of some significant feature, the striking peculiarities of the insect physiognomy!

"The gipsy who night and day for seven months goes to and fro with her brats upon her back" is the Lycosa, the Tarantula with the black stomach, the great spider of the wastes.

The larva of the great Capricornis, which gnaws the interior of old oak- trees, "leaving behind it, in the form of dry-rot, the refuse of its digestive processes," is "a scrap of intestine which eats its way as it goes."

In "that hideous lout" the Scorpion he shows us a rough epitome of the shapeless head, the truncated face of the spider.

The Tachinae, those "brazen diptera" which swarm on the sunny sand on the watch for Bembex or Philanthus, in order to establish their offspring at its expense, "are bandits clad in fustian, the head wrapped in a red handkerchief, awaiting the hour of attack!"

The Languedocian Sphex, sprawling flat upon the vine leaves, grows dizzy with the heat and frisks for very pleasure; "with its feet it taps rapidly on its resting-place, and thus produces a drumming like that of a shower of rain falling thickly on the leaves." Fabre takes a keen delight in the production of these pictures, at once so exact and lifelike; but we must not therefore suppose that his mind is incapable of the detailed descriptions necessitated by the laborious processes of minute anatomy.

Like all sciences, entomology has its uninteresting aspects when we seek to study it deeply. Yet with what interest and lucidity has Fabre succeeded in expounding the complex morphoses of the obscure and miserable larva of the Sitaris, the curious intestine of the Scarabaeus, the secret of the spawning of the weevil, and the ingenious mechanisms of the musical instruments of the Decticus and the Cicada. With what subtle art he explains the song of the cricket, how the five hundred prisms of the serrated bow set the four tympana in vibration; and how the song is sometimes muffled by a process of muting. (12/10.)

Some of the images suggested to him by the forms of animals are so beautiful that certain of his descriptions might well serve to inspire an artist, or suggest new motives of decoration in the arts of enamelling, gem-engraving, jewellery, etc.

Instead of eternally copying ancient things, or seeking inspiration in lifeless texts, why not turn our attention to the numerous and interesting motives which are scattered all around us, whose originality consists precisely in the fact that they have never yet been employed? Why torture the mind to produce more painful elaborations of awkward, frozen, poverty- stricken combinations, when Nature herself is at hand, offering the inexhaustible casket of her living marvels, full of the profoundest logic and as yet unexamined?

If the bee by means of the hexagonal prism has anticipated all the geometers in the problem of the economy of space and matter; if the Epeïra and the mollusc have invented the logarithmic spiral and its transcendent properties; if all creatures "inspired by an aesthetic which nothing escapes, achieve the beautiful" (12/11.), surely human art, which can but imitate and remember, has only to employ to its profit and transfigure into ideal images the natural beauties so profusely furnished by the Unconscious.

Modern art, influenced more especially by the subtle Japanese, is already treading this path.

What artist could ever engrave on rare metals or model in precious substances a more beautiful subject than the wonderful picture of the Tarantula offering, at the length of her extended limbs, her white sac of eggs to the sun; or the transparent nymph of the Onthophagus taurus, "as though carved from a block of crystal, with its wide snout and its enormous horns like those of the Aurochs"? (12/12.) What an undiscovered subject he might find in the nymph of the Ergatus (12/13.), with its almost incorporeal grace, as though made of "translucent ivory, like a communicant in her white veils, the arms crossed upon the breast; a living symbol of mystic resignation before the accomplishment of destiny"; or in the still more mysterious nymph of the Scarabaeus sacer, first of all "a mummy of translucent amber, maintained by its linen cerements in a hieratic pose; but soon upon this background of topaz, the head, the legs, and the thorax change to a sombre red, while the rest of the body remains white, and the nymph is slowly transfigured, assuming that majestic costume which combines the red of the cardinal's mantle with the whiteness of the sacerdotal alb."

On the other hand, what Sims or Bateman ever imagined weirder caricature than the grotesque larva of the Oniticella, with its extravagant dorsal hump; or the fantastic and alarming silhouette of the Empusa, with its scaly belly raised crozierwise and mounted on four long stilts, its pointed face, turned-up moustaches, great prominent eyes, and a "stupendous mitre": the most grotesque, the most fantastic freaks that creation can ever have evolved? (12/14.)

CHAPTER 13.

THE EPIC OF ANIMAL LIFE.

Although in his portraits and descriptions Fabre is simple and exact, and so full of natural geniality; although he can so handle his words as to render them "adequate" to reproduce the moving pictures of the tiny creatures he observes, his style touches a higher level, flashes with colour, and grows rich with imagery when he seeks to interpret the feelings which animate them: their loves, their battles, their cunning schemes, and the pursuit of their prey; all that vast drama which everywhere accompanies the travail of creation.

It is here in particular that Fabre shows us what horizons, as yet almost unexplored, what profound and inexhaustible resources science is able to offer poetry.

The breaking of egg or chrysalid is in itself a moving event; for to attain to the light is for all these creatures "a prodigious travail."

The hour of spring has sounded. At the call of the field-cricket, the herald of the spring, the germs that slumber in nymph or chrysalis have broken through their spell.

CHAPTER 13.

What haste and ingenuity are required to emerge from the natal darkness, to unwrap the swaddling-bands, to break the subterranean shells, to demolish the waxen bulkheads, to perforate the soil or to escape from prisons of silk!

The woodland bug, whose egg is a masterpiece, invents I know not what magical centre-bit, what curious piece of locksmith's work, in order to unlock its natal casket and achieve its liberty.

For days the grasshopper "butts its head against the roughness of the soil, and wars upon the pebbles; by dint of frantic wriggling it escapes from the womb of the earth, bursts its old coat, and is transfigured, opening its eyes to the light, and leaping for the first time."

The Bombyx of the pine-tree "decks its brow with points of diamond, spreads its wings, and erects its plumes, and shakes out its fleece to fly only in the darkness, to wed the same night, and to die on the morrow."

What marvellous inventions, what machinery, what incredible contrivances, "in order that a tiny fly can emerge from under ground"!

The Anthrax assumes a panoply of trepans, an assortment of gimlets and knives, harpoons and grapnels, in order to perforate its ceiling of cement; then the lugubrious black fly appears, all moist as yet with the humours of the laboratory of life, steadies itself upon its trembling legs, dries its wings, quits its suit of armour, and takes flight."

The blue-fly, buried in the depth of the sand, "cracks its barrel-shaped coffin," and splits its mask, in order to disinter itself; the head divides into two halves, between which we see emerging and disappearing by turns a monstrous tumour, which comes and goes, swells and shrivels, palpitates, labours, lunges, and retires, thus compressing and gradually undermining the sand, until at last the newborn fly emerges from the depth of the catacombs. (13/1.)

Certain young spiders, in order to emancipate themselves, to conquer space, and disperse themselves about the world, resort to an ingenious system of aviation. They gain the highest point of the thicket, and release a thread, which, seized by the wind, carries them away suspended. Each shines like a point of light against the foliage of the cypresses. There is a continuous stream of tiny passengers, leaping and descending in scattered sheaves under the caresses of the sun, like atomic projectiles, like the fountain of fire at a pyrotechnic display. What a glorious departure, what an entry into the world! Gripping its aeronautic thread, the insect ascends in apotheosis! (13/2.)

CHAPTER 13.

But if all are called all are not chosen. "How many can move only at the greatest peril under the rugged earth, proceeding from shock to shock, in the harsh womb of universal life, and, arrested by a grain of sand, succumb half-way"!

There are others whom slower metamorphoses condemn to vegetate still longer in the subterranean night, before they are permitted to assume their festival attire, and share in their turn in the gladness of creation.

Thus the Cicada is forced to labour for long gloomy years in the darkness before it can emerge from the soil. At the moment when it issues from the earth the larva, soiled with mire, "resembles a sewer-man; its eyes are whitish, nebulous, squinting, blind." Then "it clings to some twig, it splits down the back, rejects its discarded skin, drier than horny parchment, and becomes the Cigale, which is at first of a pale grass-green hue." Then,

"Half drunken with her joy, she feasts In a hail of fire";

And all day long drinks of the sugared sap of tender bark, and is silent only at night, sated with light and heat. The song, which forms part of the majestic symphony of the harvest-tide, announces merely its delight in existence. Having passed years underground, the cigale has only a month to reign, to be happy in a world of light, under the caressing sun. Judge whether the wild little cymbals can ever be loud enough "to celebrate such felicity, so well earned and so ephemeral"! (13/3.)

All sing for happiness, each after its kind, through the calm of the summer days. Their minds are intoxicated; it is their fashion of praying, of adoring, of expressing "the joys of life: a full crop and the sun on the back." Even the humble grasshopper rubs its flanks to express its joy, raises and lowers its shanks till its wing-cases squeak, and is enchanted with its own music, which it commences or terminates suddenly "according to the alternations of sun and shade." Each insect has its rhythm, strident or barely perceptible; the music of the thickets and fallows caressed by the sun, rising and falling in waves of joyful life.

The insects make merry; they hold uproarious festival; and they mate insatiably; even before forming a mutual acquaintance; in a furious rush of living, for "love is the sole joy of the animal," and "to love is to die."

Hardly unwrapped, still dusty from the strenuous labour of deliverance, "the female of the Scolia is seized by the male, who does not even give her time to wash her eyes." Having slept over a year underground, the Sitares, barely rid of their mummy-cases, taste, in the sunlight, a few minutes of love, on the very site of their re-birth; then they die. Life surges, burns, flares, sparkles, rushes "in a perpetual tide," a brief radiance between two nights.

A world of a myriad fairies fills the rustling forest: day and night it unfolds a thousand marvellous pictures; about the root of a bramble, in the shadow of an old wall, on a slope of loose soil, or in the dense thickets.

"The insect is transfigured for the nuptial ceremony; and each hopes, in its ritual, to declare its passion." Fabre had some thought of writing the Golden Book of their bridals and their wedding festivals (13/4.); the Kamasutra of their feasts and rules of love; and with what art, at once frank and reserved, has he here and there handled this wonderful theme! In the radiant garden of delight, where no detail of truth is omitted, but where nothing shocks us, Fabre reveals himself as he is in his conversation; evading the subject where it takes a licentious turn; fundamentally chaste and extremely reserved.

At the foot of the rocks the Psyche "appears in the balcony of her boudoir, in the rays of the caressing sun; lying on the cloudy softness of an incomparable eider-down." She awaits the visit of the spouse, "the gentle Bombyx," who, for the ceremony, "has donned his feathery plumes and his mantle of black velvet." "If he is late in coming, the female grows impatient; then she herself makes the advances, and sets forth in search of her mate."

Drawn by the same voluptuous and overwhelming force, the cricket ventures to leave his burrow. Adorned "in his fairest attire, black jacket, more beauteous than satin, with a stripe of carmine on the thigh," he wanders through the wild herbage, "by the discreet glimmer of twilight," until he reaches the distant lodging of the beloved. There at last he arrives "upon the sanded walk, the court of honour that precedes the entry." But already the place is occupied by another aspirant. Then the two rivals fall upon one another, biting one another's heads, "until it ends by the retreat of the weaker, whom the victor insults by a bravura cry." The happy champion bridles, assuming a proud air, as of one who knows himself a handsome fellow, before the fair one, who feigns to hide herself behind her tuft of aphyllantus, all covered with azure flowers. "With a gesture of a fore-limb he passes one of his antennae through his mandibles as though to curl it; with his long-spurred, red-striped legs he shuffles with impatience; he kicks the empty air; but emotion renders him mute." (13/5.)

In the foliage of the ash-tree the lover of the female Cantharis thrashes his companion, who makes herself as small as she can, hiding her head in her bosom; he bangs her with his fists, buffets her with his abdomen, "subjects her to an erotic storm, a rain of blows"; then, with his arms crossed, he remains a moment motionless and trembling; finally, seizing both antennae of the desired one, he forces her to raise her head "like a cavalier proudly seated on horse and holding the reins in his hands."

The Osmiae "reply by a click of the jaws to the advances of their lovers, who recoil, and then, doubtless to make themselves more valiant, they also execute a ferocious mandibular grimace. With this byplay of the jaws and their menacing gestures of the head in the empty air the lovers have the air of intending to eat one another." Thus they preface their bridals by displays of gallantry, recalling the ancient betrothal customs of which Rabelais speaks; the pretenders were cuffed and derided and threatened with a hearty pummelling. (13/6.)

On the arid hillsides, where the doubtful rays of the moon pierce the storm-clouds and illumine the sultry atmosphere, the pale scorpions, with short-sighted eyes, hideous monsters with misshapen heads, "display their strange faces, and two by two, hand in hand, stalk in measured paces amid the tufts of lavender. How tell their joys, their ecstasies, that no human language can express...!" (13/7.)

However, the glow-worm, to guide the lover, lights its beacon "like a spark fallen from the full moon"; but

CHAPTER 13. 272

"presently the light grows feebler, and fades to a discreet nightlight, while all around the host of nocturnal creatures, delayed in their affairs, murmur the general epithalamium." (13/8.)

But their happy time is soon over; tragedy is about to follow idyll.

One must live, and "the intestine rules the world."

All creatures that fill the world are incessantly conflicting, and one lives only at the cost of another.

On the other hand, in order that the coming generations may see the light, the present generations must think of the preservation of the young. "Perish all the rest provided the brood flourish!" And in the depth of burrows the future larvae who live only for their stomachs, "little ogres, greedy of living flesh," must have their prey.

To hunger and maternity let us also add love, which "rules the world by conflict."

Such are the components of the "struggle for existence," such as Fabre has described it, but with no other motive than to describe what he has observed and seen. Such are the ordinary themes of the grandiose battles which he has scattered through his narratives, and never did circus or arena offer more thrilling spectacles; no jungle ever hid more moving combats in its thickets."

"Each has its ruses of war, its methods of attack, its methods of killing."

What tactics--"studied, scientific, worthy of the athletes of the ancient palaestra"--are those which the Sphex employs to paralyse the Cricket and the Cerceris to capture the Cleona, to secure them in a suitable place, so as to operate on them more surely and at leisure!

Beside these master paralysers, so expert in the art of dealing slow death, there are those which, with a precision no less scholarly, kill and wither their victims at a single stroke, and without leaving a trace: "true practitioners in crime."

On the rock-rose bushes, with their great pink flowers, "the pretty Thomisus, the little crab-spider, clad in satin," watches for the domestic bee, and suddenly kills it, seizing the back of the head, while the Philanthus, also seizing it by the head, plunges its sting under the chin, neither too high nor too low, but "exactly in the narrow joint of the neck," for both insects know that in this limited spot, in which is concentrated a small nervous mass, something

CHAPTER 13.

like a brain, is "the weak point, most vulnerable of all," the fault in the cuirass, the vital centre. Others, like the Araneidae, intoxicate their prey, and their subtle bite, "which resembles a kiss," in whatever part of the body it is applied, "produces almost immediately a gradual swoon."

Thus the great hairy Bourdon, in the course of its peregrinations across the wastes of thyme, sometimes foolishly strays into the lair of the Tarantula, whose eyes glimmer like jewels at the back of his den. Hardly has the insect disappeared underground than a sort of shrill rattling is heard, a "true death-song," immediately followed by the completest silence. "Only a moment, and the unfortunate creature is absolutely dead, proboscis outstretched and limbs relaxed. The bite of the rattlesnake would not produce a more sudden paralysis."

The terrible spider "crouching on the battlements of his castle, his heavy belly in the sun, attentive to the slightest rustling, leaps upon whatever passes, fly or Libellula, and with a single stroke strangles his victim, and drains its body, drinking the warm blood."

"To dislodge him from his keep needs all the cunning strategy of the Pompilus; a terrible duel, a hand-to-hand combat, stupendous, truly epic, in which the subtle address and the ingenious audacity of the winged insect eventually triumph over the dreadful spider and his poisoned fangs." (13/9.)

On the pink heather "the timid spider of the thickets suspends by ethereal cables the branching whorl of his snare, which the tears of the night have turned into chaplets of jewels...The magical jewellery sparkles in the sun, attracting mosquitoes and butterflies; but whosoever approaches too closely perishes, a victim of curiosity." Above the funnel is the trap, "a chaos of springs, a forest of cordage; like the rigging of a ship dismembered by the tempest. The desperate creature struggles in the shrouds of the rigging, then falls into the gloomy slaughter-house where the spider lurks ready to bleed his prey."

Death is everywhere.

Each crevice of bark, each shadow of a leaf, conceals a hunter armed with a deadly weapon, all his senses on the alert. Everywhere are teeth, fangs, talons, stings, pincers, and scythes.

Leaping in the long grasses, the Decticus with the ivory face "crunches the heads of grasshoppers in his mandibles."

CHAPTER 13.

A ferocious creature, the grub of the Hemerobius, disembowels plant-lice, making of their skins a battle-dress, covering its back with the eviscerated victims, "as the Red Indian ties about his loins the tresses of his scalped enemies."

Caterpillars are surrounded by the implacable voracity of the Carabidae:

"The furry skins are gaping with wounds; their contents escape in knots of entrails, bright green with their aliment, the needles of the pine-tree; the caterpillars writhe, struggling with loop-like movements, gripping the sand with their feet, dribbling and gnashing their mandibles. Those as yet unwounded are digging desperately in the attempt to escape underground. Not one succeeds. They are scarcely half buried before some beetle runs to them and destroys them by an eviscerating wound."

At the centre of its net, which seems "woven of moonbeams," in the midst of its snare, a glutinous trap of infernal ingenuity, or hidden at a distance in its cabin of green leaves, the Epeïra fasciata waits and watches for its prey. Let the terrible hornet, or the Libellula auripennis, flying from stem to stem, fall into the limed snare; the insect struggles, endeavours to unwind itself; the net trembles violently as though it would be torn from its cables. Immediately the spider darts forward, running boldly to the intruder. With rapid gestures the two hinder limbs weave a winding- sheet of silk as they rotate the victim in order to enshroud it...The ancient Retiarius, condemned to meet a powerful beast of prey, appeared in the arena with a net of cordage lying upon his left shoulder; the animal sprang upon him; the man, with a sudden throw, caught it in the meshes; a stroke of the trident despatched it. Similarly the Epeïra throws its web, and when there is no longer any movement under the white shroud the spider draws closer; its venomous fangs perform the office of the trident. (13/10.)

The Praying Mantis, that demoniac creature which alone among the insects turns its head to gaze, "whose pious airs conceal the most atrocious habits," remains on the watch, motionless, for hours at a time. Let a great grasshopper chance to come by: the Mantis follows it with its glance, glides between the leaves, and suddenly rises up before it; "and then assumes its spectral pose, which terrifies and fascinates the prey; the wing-covers open, the wings spring to their full width, forming a vast pyramid which dominates the back; a sort of swishing sound is heard, like the hiss of a startled adder; the murderous fore-limbs open to their full extent, forming a cross with the body, and exhibiting the axillae ornamented with eyes vaguely resembling those of the peacock's tail, part of the panoply of war, concealed upon ordinary occasions. These are only exhibited when the creature makes itself terrible and superb for battle. Then the two grappling-hooks are thrown; the fangs strike, the double scythes close together and hold the victim as in a vice." (13/11.)

There is no peace; night falls and the horrible conflict continues in the darkness. Atrocious struggles, merciless duels, fill the summer nights. On the stems of the long grasses, beside the furrows, the glow-worm "anaethetizes the snail," instilling into it its venom, which stupefies and produces sleep, in order to immobilize its prey before devouring it.

Having chorused their joy all the day long in the sunshine, in the evening the Cicadae fall asleep among the olives and the lofty plane-trees. But suddenly there is a sound as of a cry of anguish, short and strident; it is the despairing lamentation of the cicada, surprised in repose by the green grasshopper, that ardent hunter of the night, which leaps upon the cicada, seizes it by the flank, and devours the contents of the stomach. After the orgy of music comes night and assassination.

Such is the gloomy epic which goes forward among the flowers, amidst the foliage, under the shadowy boughs, and on the dusty fallows. Such are the sights that nature offers amid the profound peace of the fields, behind the flowering of the sudden spring-tide and the splendours of the summer. These murders, these assassinations are committed in a mute and silent world, but "the ear of the mind" seems to hear

"A tiger's rage and cries as of a lion Roaring remotely through this pigmy world."

Was it to these thrilling revelations that Victor Hugo intended to apply these so wonderfully appropriate lines? Was it he who bestowed upon Fabre, according to a poetic tradition, the name of "the Homer of the insects," which fits him so marvellously well?

It is possible, although Fabre himself can cite no evidence to support these suggestions; but let us respect the legend, simply because it is charming, and because it adds an exact and picturesque touch to the portrait of Fabre.

In this drama of a myriad scenes, in which the little actors in their rustic stage play each in his turn their parts at the mercy of occasion and the hazard of encounter, the humblest creatures are personages of importance.

Like the human comedy, this also has its characters privileged by birth, clothed in purple, dazzling with embroidery, "adorned with lofty plumes," who strut pretentiously; "its idle rich," covered with robes of gold of rustling splendour, who display their diamonds, their topazes and their sapphires; who gleam with fire and shine like mirrors, magnificent of mien; but their brains are "dense, heavy, inept, without imagination, without ingenuity, deprived of all common sense, knowing no other anxiety than to drink in the sunlight at the heart of a rose or to sleep off their draughts in the shadow of a leaf.

Those who labour, on the contrary, do not attract the eye, and the most obscure are often the most interesting. Necessitous poverty has educated and formed them, has excited in them "feats of invention," unsuspected talents, original industries; a thousand curious and unexpected callings, and no subject of poetry equals in interest the detailed history of one of these tiny creatures, by which we pass without observing them, amid the stones, the brambles, and the dead leaves. It is these above all that add an original and epic note to the vast symphony of the world.

But death also has its poetry. Its shadowy domains hold lessons no less magnificent, and the most putrid carrion is to Fabre a "tabernacle" in which a divine comedy is enacted.

CHAPTER 13.

The ant, that "ardent filibuster, comes first, and commences to dissect it piecemeal."

The Necrophori "exhaling the odour of musk, and bearing red pompons at the end of their antennae," are "transcendent alchemists."

The Sarcophagi, or grey flesh flies, "with red bloodshot eyes, and the stony gaze of a knacker"; the Saprinidae, "with bodies of polished ebony like pearls of jet"; the Silpha aplata, with large and sombre wing-cases in mourning; the shiny slow-trotting Horn-beetle; the Dermestes, "powdered with snow beneath the stomach"; the slender Staphylinus; the whole fauna of the corpse, the whole horde of artisans of death, "intoxicating themselves with purulence, probing, excavating, mangling, dissecting, transmuting, and stamping out infection."

Fabre gives a curious exposition of "that strange art" by which the grub of the grey bot-fly, the vulgar maggot, by means of a subtle pepsine, disintegrates and liquefies solid matter; and it is because this singular solvent has no effect upon the epidermis that the fly, in its wisdom, chooses by preference the mucous membranes, the corner of the eye, the entrance of the nostrils, the borders of the lips, the live flesh of wounds, there to deposit its eggs.

CHAPTER 13.

With what penetration this original mind has analysed "the operation of the crucible in which all things are fused that they may recommence" and has expounded the marvellous lesson which is revealed by decomposition and putridity!

CHAPTER 14.

PARALLEL LIVES.

We have now seen what entomology becomes in the hands of the admirable Fabre. The vast poem of creation has never had a more familiar and luminous interpreter, and you will nowhere find other work like his.

How far he outstrips Buffon and his descriptions of animals--so general, so vague, so impersonal--his records unreliable and his entire erudition of a second-hand quality!

It is with Réaumur that we are first of all tempted to compare him; and some have chosen to see in him only one who has continued Réaumur's work. In reality he has eagerly read Réaumur, although at heart he does not really enjoy his writings; he has drunk from this fruitful source, but he owes him no part of his own rich harvest.

But there are many affinities between them; they have many traits in common, despite the points of difference between them.

The illustrious son of Rochelle was born, like Fabre, with a love of all natural things, and before attacking the myriad problems of physics and natural history, wherein he was to shine by so many curious discoveries, he also had prepared himself by a profound study of mathematics.

Luckier than Fabre, however, Réaumur enjoyed not only the advantages of birth, but all the material conditions necessary to his ardent intellectual activity. Fortune overwhelmed her favourite with gifts, and played no small part in his glory by enabling him, from an early age, to profit by his leisure and to give a free rein to his ruling passions. He was no less modest than the sage of Sérignan; self-effacing before others, says one of his biographers, so that they were never made to feel his superiority. (14/1.)

In the midst of the beautiful and spacious gardens at the end of the Faubourg Saint-Antoine, where he finally made his home, he also contrived to create for himself a Harmas after his own heart.

It was there that in the as yet virgin domain of entomology he unravelled the riddle of the marvellous republic of the bees, and was able to expound and interpret a large number of those tiny lives which every one had hitherto despised, and which indeed they continued to despise until the days of Fabre, or at least regarded as absolutely unimportant. He was the first to venture to suspect their connection with much "that most nearly concerns us," or to

CHAPTER 14.

point out "all the singular conclusions" which may be drawn therefrom. (14/2.)

How many details he has enshrined in his interesting "Memoirs," and how many facts we may glean from this great master! He, like Fabre, had the gift of charming a great number of his contemporaries. Tremblay, Bonnet, and de Geer owed their vocations to Réaumur, not to speak of Huber, whose genius he inspired.

A physicist before all, and accustomed to delicate and meticulous though comparatively simple tasks, he had admirably foreseen the extraordinary complication of these inquiries; so much so that, with the modesty of the true scientist that he was, he regarded his own studies, even the most substantial, as mere indications, intended to point the way to those that followed him.

As methodical, in short, as the author of the "Souvenirs," the scrupulous Réaumur wrote nothing that he himself had not proved or verified with the greatest care; and we may be sure that all that he records of his personal and immediate observations he has really seen with his own eyes.

In the wilderness of error he had, like Fabre, an infallible compass in his extraordinary common sense; and, equally skilled in extracting from the false the little particle of truth which it often contains, he was no less fond of listening at the gate of legends, of tracing the source of traditions; rightly considering that before deriding them as old-wives' tales we should first probe in all directions into their origin and foundation. (14/3.)

He was also tempted to experiment, and he well knew that in such problems as those he attacked observation alone is often powerless to reveal anything. It is enough to recall here one of the most promising and unexpected of the discoveries which resulted from his experiments. Réaumur was the first to conceive the ingenious idea of retarding the hatching of insects' eggs by exposing them to cold, thus anticipating the application of cold to animal life and the discoveries of Charles Tellier, whose more illustrious forerunner he was; at the same time he discovered the secret of prolonging, in a similar fashion, the larval existence of chrysalids during a space of time infinitely superior to that of their normal cycle; and what is more, he succeeded in making them live a lethargic life for years and even for a long term of years, thus repeating at will the miracle of the Seven Sleepers. (14/4.)

Too much occupied, however, with the smaller aspect of things, he had not the art of forcing Nature to speak, and in the province of psychical aptitudes he was barely able to rise above the facts.

As he was powerless to enter into real communion with the tiny creatures which he observed, although his observations were conducted with religious admiration; as he saw always only the outside of things, like a physicist rather than a poet or psychologist, he contented himself with noting the functioning of their organs, their methods of work, their properties, and the changes which they undergo; he did not interpret their actions. The mystery of the life which quivers within and around them eludes him. This is why his books are such dry reading. He is like a bright garden full of rare plants; but it is a monotonous garden, without life or art, without distant vistas or wide perspectives. His works are somewhat diffuse and full of repetitions; entire monographs, almost whole volumes, are devoted to describing the emerging of a butterfly; but they form part of the library of the curious lover of nature; they are consulted with interest, and will always be referred to, but it cannot be said that they are read.

After Réaumur, according to the dictum of the great Latreille, entomology was confined to a wearisome and interminable nomenclature, and if we except the Hubers, two unparalleled observers, although limited and circumscribed, the only writer who filled the interregnum between Réaumur and Fabre was Léon Dufour.

CHAPTER 14.

In the quiet little town whither he went to succeed his father, this military surgeon, turned country doctor, lived a busy and useful life.

While occupied with his humble patients, whom he preferred to regard merely as an interesting clinic, and while keeping the daily record of his medical observations, he felt irresistibly drawn "to ferret in all the holes and corners of the soil, to turn over every stone, large or small; to shrink from no fatigue, no difficulty; to scale the highest peaks, the steepest cliffs, to brave a thousand dangers, in order to discover an insect or a plant. (14/5.)

A disciple of Latreille, he shone above all as an impassioned descriptive writer.

No one was more skilled in determining a species, in dissecting the head of a fly or the entrails of a grub, and no spectacle in the world was for him so fascinating as the triple life of the insect; those magical metamorphoses, which he justly considered as one of the most astonishing phenomena in creation. (14/6.)

He saw further than Réaumur, and burned with the same fire as Fabre, for he also had the makings of a great poet. His curiosity had assembled enormous collections, but he considered, as Fabre considered, that collecting is "only the barren contemplation of a vast ossuary which speaks only to the eyes, and not to the mind or imagination," and that the true history of insects should be that of their habits, their industries, their battles, their loves, and their private and social life; that one must "search everywhere, on the ground, under the soil, in the waters, in the air, under the bark of trees, in the depth of the woods, in the sands of the desert, and even on and in the bodies of animals."

Was not this in reality the ambitious programme which Fabre was later to propose to himself when he entered into his Harmas and founded his living laboratory of entomology; he also having set himself as his exclusive object the study of "the insects, the habits of life, the labours, the struggles and the propagation of this little world, which agriculture and philosophy should closely consider"? (14/7.)

Dufour also had admirably grasped the place of the insect in the general harmony of the universe, and he clearly perceived that parasitism, that imbrication of mutually usurping lives, is "a law of equilibration, whose object is to set a limit to the excessive multiplication of individuals of the same type," that the parasites are predestined to an imprescriptible mission, and that this mysterious law "defies all explanation."

On the other hand, he did not become very intimate with these tiny peoples; his attention was dispersed over too many points; perhaps he was fundamentally incapable of concentrating himself for a long period upon a circumscribed object; perhaps he lacked that first condition of genius, patience, so essential to such researches: although he enriched science by an infinite multitude of precious facts and has recorded a quantity of details concerning the habits of insects, he did not succeed in representing any one of these innumerable little minds. He had an intense feeling for nature, but he was not able to interpret it, and his immense volume of work, scattered through nearly three hundred monographs, remains ineffective.

Let us compare with his work the vast epic of the "Souvenirs." We become familiar with the whole life of the least insect, and all its unending related circumstances; we obtain sudden glimpses of insight into our own organization, with its abysses and its lacunae, and also into those rich provinces or faculties which we are only beginning to suspect in the depths of our unconscious activity.

In the evening twilight, after the vast andante of the cicadae is hushed, at the hour when the shining glow-worms "light their blue fires," and the "pale Italian cricket, delirious with its nocturnal madness, chirrups among the rosemary thickets," while in the distance sounds the melodious tinkle of the bell-ringer frogs, replying from one hiding-place to another, the old master shows us that profound and mysterious magic with which matter is endowed by the faintest glimmer of life.

He shows us the intimate connection of things, the universal harmony which so intimately allies all creatures; and he shows us also that everywhere and all around us, in the smallest object, poetry exists like a hidden flame, if only we know how to seek it.

And in revealing so many marvellous energies in even the lowest creatures, he helps us to divine the infinity of phenomena still unguessed-at, which the subtlety of the unknowable force which thrills through the whole universe hides from us under the most trivial appearances.

For he has not told everything; this incommensurable region, which had hitherto remained unworked, is far from being exhausted.

How many unknown and hidden things are still left to be gleaned! There will be a harvest for all. Remember that "even the humblest species either has no history, or the little that has been written concerning it calls for serious revision" (14/8.); that a single bush, such as the bramble, suffices to rear more than fifty species of insects, and that each species, according to the just observation of Réaumur, "has its habits, its tricks of cunning, its customs, its industries, its art, its architecture, its different instincts, and its individual genius."

What a stupendous alphabet to decipher, of which we have as yet only commenced to read the first few letters! When we are able to read it almost entirely, when observers are more numerous and have concerted their efforts, mutually illuminating, completing and correcting one another, then, and then only, we shall succeed, if not in resolving some of those high problems which have never ceased to interest mankind, at least in seizing some reflected knowledge of ourselves, and in seeing a little farther into the kingdom of the mind.

CHAPTER 15.

THE EVENINGS AT SÉRIGNAN.

But it will doubtless be long before a new Fabre will resume, with the same heroic ardour, the life of solitary labour, varied only by a few austere recreations.

Rising at six o'clock, he would first of all pace the tiles of his kitchen, breakfast in hand; so imperious in him was the need of action, if his mind was to work successfully, that even at this moment of morning meditation his body must already be in movement. Then, after many turns among the bushes of the enclosure, all irised with drops of dew which were already evaporating, he went straight to his cell: that is, to the silence of his laboratory.

There, in unsociable silence, invisible to all, he worked hard and steadily until noon; pursuing an observation or carrying out some experiment, or recording what he saw or what he had seen the day before, or re-drafting his records in their final form.

How many who have come hither to knock upon the door in these morning hours, or to ring at the little gate, silent as the tomb, which gives upon the private path frequented only by foot-passengers on their way to the fields, have undertaken a fruitless journey! But without such discipline would it have been possible to accomplish such a task as his?

At last he would leave his workroom; jaded, exhausted by the excessive intensity of his work, "face pale and features drawn." (15/1.)

Now he is "at leisure: the half-day is over" (15/2.); and he can satisfy his immense need not of repose, but of relaxation and distraction in less severe occupations; for he is never at any time nor anywhere inactive; incessantly making notes, with little stumps of pencil which he carries about in his pockets, and on the first scrap of paper that comes to hand, of all that passes through his mind. Those eternal afternoons, which usually, in the depth of the French provinces, prove so dull and wearisome, seem short enough to him. Now he will halt before his plants, now stoop to the ground, the better to observe a passing insect; always in search of some fresh subject of study; or now bending over his microscope. (15/3.) Then he undertakes, for his later-born children at Sérignan, the duties which he formerly performed for the elder family at Orange: he teaches them himself; he has much to do with them, for their sake and for his own as well, for he is jealous of possessing them, and he regrets parting with them. They too have their tasks arranged in advance.

CHAPTER 15.

They are his assistants, his appointed collaborators, who keep and relieve guard, undertaking, in his absence, some observation already in hand, so that no detail may be lost, no incident of the story that unrolls itself sometimes with exasperating slowness beneath the bell-covers of the laboratory or on some bush in the garden. He inspires the whole household with the fire of his own genius, and all those about him are almost as interested as he.

At home, in the house, always wearing his eternal felt hat, and absorbed in meditation, he speaks little, holding that every word should have its object, and only employing a term when he has tested its weight and meaning. Silence at mealtimes again is a rule that no one of his household would infringe. But he unbends his brow when he receives a friend at his hospitable table, where but lately his smiling wife would sit, full of little attentions for him. (15/4.)

Frugal in all respects, he barely touches the dishes before him; avoiding all meats, and saving himself wholly for the fruits; for is not man naturally frugivorous, by his teeth, his stomach, and his bowels? Certain dishes repel him, for reasons of sentiment rather than through any real disgust; such as paté de foie gras, which reminds him too forcibly of the so cruelly tortured goose; such cruelty is too high a price to pay for a mere greasy mouthful. (15/5.) On the other hand, he drinks wine with pleasure, the harsh, rough "wine of the country" of the plains of Sérignan. He is also well able to appreciate good things and appetizing cookery; no one ever had a finer palate; but he is happiest in seeing others appreciate the pleasures of the table. Witness that breakfast worthy of Gargantua, which he himself organized in honour of his guests, whom he had invited to an excursion over the Ventoux Alp; where he seems expressly to have commanded "that all should come in shoals." What a tinkling of bottles, what piles of bread! There are green olives "flowing with brine," black olives "seasoned with oil," sausages of Arles "with rosy flesh, marbled with cubes of fat and whole peppercorns," legs of mutton stuffed with garlic "to dull the keen edge of hunger"; chickens "to amuse the molars"; melons of Cavaillon too, with white pulp, not forgetting those with orange pulp, and to crown the feast those little cheeses, so delightfully flavoured, peculiar to Mont Ventoux, "spiced with mountain herbs," which melt in the mouth. (15/6.)

But his greatest pleasure is his pipe; a briar, which in absence of mind he is always allowing to go out, and always relighting.

Respectful of all traditions, he has kept up the observance of old customs; no Christmas Eve has ever been passed under the roof of his Harmas without the consecrated meats upon the table; the heart of celery, the nougat of almonds, the dish of snails, and the savoury-smelling turkey. Then, stuck into the Christmas bread (15/7.), the sprigs of holly, the verbouisset, the sacred bush whose little starry flowers and coral berries, growing amid evergreen leaves, affirm the eternal rebirth of indestructible nature.

At Sérignan Fabre is little known and little appreciated. To tell the truth, folk regard him as eccentric; they have often surprised him in the country lying on his stomach in the middle of a field, or kneeling on the ground, a magnifying glass in hand, observing a fly or some one of those insignificant creatures in which no sane person would deign to be interested.

How should they know him, since he never goes into the village? When he did once venture thither to visit his friend Charrasse, the schoolmaster, his appearance was an event of which every one had something to say, so greatly did it astonish the inhabitants. (15/8.)

Yet he never hesitates to place his knowledge at the service of all, and welcomes with courtesy the rare pilgrims in whom a genuine regard is visible, although he is always careful never to make them feel his own superiority; but he very quickly dismisses, sometimes a trifle hastily, those who are merely indiscreet or importunate; pedantic and ignorant persons he judges instantaneously with his piercing eyes; with such people he cannot emerge from his slightly gloomy reserve; he shuts himself up like the snail, which, annoyed by some displeasing object, retires into its shell, and remains silent in their presence.

Professors come to consult him: asking his advice as to their programmes of instruction, or begging him to resolve some difficult problem or decide some especially vexed question; and his explanations are so simple, so clear, so logical that they are astonished at their own lack of comprehension and their embarrassment. (15/9.)

But there are few who venture within the walls of that enclosure, which seems to shut out all the temptations of the outer world; the only intimate visitors to the Harmas are the village schoolmaster--first Laurent, then Louis Charrasse (15/10.), and later Jullian--and a blind man, Marius.

This latter lost his sight at the age of twenty. Then, to earn a living, he began to make and repair chairs, and in his misfortune, although blind and extremely poor, he kept a calm and contented mind.

Fabre had discovered the sage and the blind man on his arrival at Sérignan, and also Favier (15/11.), "that other

CHAPTER 15.

native, whose jovial spirit was so prompt to respond, and who helped to dig up the Harmas; to set up the planks and tiles of the little kitchen-garden; a rude task, since this scrap of uncultivated ground was then but a terrible desert of pebbles." To Favier fell the care of the flowers, for the new owner was a great lover of flowers. Potted plants, sometimes of rare species, were already, as to-day, crowded in rows upon the terrace before the house, where all the summer they formed a sort of vestibule in the open air, on either side of the entrance; and these Fabre never ceased to watch over with constant and meticulous care. Both spoke the same language, and the words they exchanged were born of a like philosophy; for Favier also loved nature in his own way, and at heart was an artist; and when, after the day's work, sitting "on the high stone of the kitchen hearth, where round logs of green oak were blazing," he would evoke, in his picturesque and figurative language, the memories of an old campaigner, he charmed all the household and the evening seemed to pass with strange rapidity.

When this precious servant and boon companion had disappeared, after two years of digging, sowing, weeding, and hoeing, all was ready; the frame was completed and the work could be commenced. It was then that Marius became the master's appointed collaborator, and it is he who now constructs his apparatus, his experimental cages;

stuffs his birds, helps to ransack the soil, and shades him with an umbrella while he watches under the burning sun. Marius cannot see, but so intimate is his communion with his master, so keen his enthusiasm for all that Fabre does, that he follows in his mind's eye, and as though he could actually see them, all the doings at which he assists, and whose inward reflection lights up his wondering countenance.

Marius was not only rich in feeling and the gift of inner vision; he had also a marvellously correct ear. He was a member of the "Fanfare" of Sérignan, in which he played the big drum, and there was no one like him for keeping perfect time and for bringing out the clash of the cymbals.

Charrasse was no less fervent a disciple; he worshipped science and all beautiful things; and he could even conceive a noble passion for his exhausting trade of school-teaching.

Like Marius, he ate "a bitter bread"; and Fabre would get on with them all the better in that they, like himself, had lived a difficult life. "Man is like the medlar," he liked to tell them; "he is worth nothing until he has ripened a long time in the attic, on the straw."

CHAPTER 15.

"L'homme est comme la nèfle, il n'est rien qui vaille S'il n'a mûri longtemps, au grenier, sur la paille."

These humble companions afforded him the simple conversation which he likes so well; so natural, and so full of sympathy and common sense. They customarily spent Thursday and Sunday afternoons at the Harmas; but these beloved disciples might call at any hour; the master always welcomed them, even in the morning, even when he was entirely absorbed in his work and could not bear any one about him. They were his circle, his academy; he would read them the last chapter written in the morning; he shared his latest discoveries with them; he did not fear to ask advice of their "fertile ignorance." (15/12.)

Charrasse was a "Félibre," versed in all the secrets of the Provençal idiom, of which he knew all the popular terms, the typical expressions and turns of speech; and Fabre loved to consult him, to read some charming verses which he had just discovered, or to recite some delightful rustic poem with which he had just been inspired; for in such occupations he found one of his favourite relaxations, giving free vent to his fancy, a loose rein to the poet that dwells within him. These poems the piety of his brother has preserved in the collection entitled "Oubreto." It is at such a moment that one should see his black eyes, full of fire; his power of mimicry and expression, his impassioned features, lit up by inspiration, truly idealized, almost transfigured, are at such times a thing to be remembered.

Sometimes, again, in the shadow of the planes, on summer afternoons, when the cigales were falling silent; or in the winter, before the blazing fireplace, in that dining-room on the ground floor in which he welcomed his visitors; when out of doors the mistral was roaring and raging, or the rain clattering on the panes, the little circle was enlarged by certain new- comers, his nephews, nieces, a few intimates, of whom, a little later, I myself was often one. At such times his humour and imagination were given full play, and it was truly a rare pleasure to sit there, sipping a glass of mulled wine, during those delightful and earnest hours; to taste the charm of his smiling philosophy, his picturesque conversation, full of exact ideas, all the more profound in that they were founded on experience and pointed or adorned by proverbs, adages, and anecdotes. Thanks to the daily reading of the "Temps," which one of his friends regularly sends him, Fabre is in touch with all the ideas of the day, and expresses his judgment of them; for example, he does not conceal his scepticism with regard to certain modern inventions, such as the aeroplane, whose novelty rather disturbs his mind, and whose practical bearing seems to him to be on the whole somewhat limited.

CHAPTER 15.

Thus even the most recent incidents find their way into the solitude of the Harmas and help to sustain the conversation.

"The first time we resume our Sérignan evenings," he wrote to his nephew on the morrow of one of these intimate gatherings, "we will have a little chat about your Justinian, whom the recent drama of "Théodora" has just made the fashion. Do you know the history of that terrible hussy and her stupid husband? Perhaps not entirely; it is a treat I am keeping for you." (15/13.)

The only subject which is hardly ever mentioned during these evenings at Sérignan is politics, although Fabre, strange as it may seem, was one year appointed to sit on the municipal council.

The son of peasants, who has emerged from the people yet has always remained a peasant, has too keen a sense of injustice not to be a democrat; and how many young men has he not taught to emancipate themselves by knowledge? But above all he is proud of being a Frenchman; his mind, so lucid, so logical, which has never gone abroad in search of its own inspirations, and has never been influenced by any but those old French masters, François Dufour and Réaumur, and the old French classics, has always felt an instinctive repugnance, which it has never been able to overcome, for all those ideas which some are surreptitiously seeking to put forward in our midst in favour of some foreign trade-mark.

Although his visit to the court of Napoleon III left him with a rather sympathetic idea of the Emperor, whose gentle, dreamy appearance he still likes to recall, he detested the Empire and the "brigand's trick" which established it.

On the day of the proclamation of the Republic he was seen in the streets of Avignon in company with some of his pupils. He was agreeably surprised at the turn events had taken, and delighted by the unforeseen result of the war.

A spirit as proud and independent as his was naturally the enemy of any species of servitude. State socialism of the equalitarian and communistic kind was to him no less horrifying. Was not Nature at hand, always to remind him of her eternal lessons?

"Equality, a magnificent political label, but scarcely more! Where is it, this equality? In our societies shall we find even two persons exactly equal in vigour, health, intelligence, capacity for work, foresight, and so many other gifts which are the great factors of prosperity?...A single note does not make a harmony: we must have dissimilar notes; discords even, which, by their harshness, give value to the concords; human societies are harmonious only thus, by the concourse of dissimilarities." (15/14.)

And what a puerile Utopia, what a disappointing illusion is that of communism! Let us see under what conditions, at the price of what sacrifices, nature here and there realizes it.

Among the bees "twenty thousand renounce maternity and devote themselves to celibacy to raise the prodigious family of a single mother."

Among the ants, the wasps, the termites "thousands and thousands remain incomplete and become humble auxiliaries of a few who are sexually gifted."

Would you by chance reduce man to the life of the Processional caterpillars, content to nibble the pine-needles among which they live, and which, satisfied to march continually along the same tracks, find within reach an abundant, easy, and idle subsistence? All have the same size, the same strength, the same aptitudes. No initiative. "What one does the others do, with equal zeal, neither better nor worse." On the other hand, there is "no sex, no love." And what would be a society in which there was no work done for pleasure and from which love and the family were banished? What would be the effect upon its progress, its welfare, its happiness? Would not all that make the charm of life disappear for good? However imperfect our present society may be, however mysterious its destinies, it is not in socialism that Fabre foresees the perfection of future humanity, for to him the true humanity does not as yet exist; it is making its way, it is slowly progressing, and in this evolution he wishes with all his heart to believe. Modern humanity is as yet only a shapeless grimacing caricature, and its life is like a play written by madmen and played by drunken actors; according to those profound words of the great poet, with which his mind is in some sort imbued; which he often repeats, and which he has transcribed at the head of one of his last records as an epigraph and a constant reminder.

And you who groan over the distressing problem of depopulation, lend an ear to the lesson of the Copris, "which trebles its customary batch of offspring in times of abundance, and in times of dearth imitates the artisan of the city who has only just enough to live on, or the bourgeois, whose numerous wants are more and more costly to satisfy, limiting the number of its offspring lest they should go in want, often reducing the number of its children to a single one." (15/15.)

CHAPTER 15.

Instead of running after so many false appearances and false pleasures, learn to return to simpler tastes, to more rustic manners; free yourselves from a mass of factitious needs; steep yourself anew in the antique sobriety, whose desires were sager; return to the fields, the source of abundance, and the earth, the eternal foster-mother!

And in this appeal to return to nature, which perhaps since the time of Rousseau has never been worded so eloquently, Fabre has in view if not the strong, the predestined, who are called elsewhere, and who are actuated by the sense of great tasks to be performed, at least all those of rural origin, all those for whom the love of the family, the daily task, and a peaceful heart are really the great things of life, the things that count, the things that suffice.

He himself, although he was one of the strong, did not care to break any of the ties that bound him to his origins. Like the Osmia, "which retains a tenacious memory of its home," the beloved village of his childhood has never been effaced from his memory, and for a long time the desire to leave his bones there haunted him. His mind often returned to it; he thought that there, better than anywhere else, he would find peace; that it would please him to wander among the rocks, the trees, the stones which he had so loved, in the old days, and that all these things would

CHAPTER 15. 308

recognize him too.

One day, however, when I was begging him to make up his mind on this point- -it was one of those peaceful evenings which are troubled under the plane- trees only by the tinkling of the fountain--he confided to me that his beloved Sérignan had at last, in his secret preferences, obliterated the old longing. As he advanced in life, in fact, although he never forgot his rude natal countryside, he felt that new links were daily binding him more closely to those heaths and mountains on which his heart had been so often thrilled with the intense joy of discovery, and that it was indeed in this soil, to him so full of delight, amid its beautiful hymenoptera and scarabaei, that he would wish to be buried.

Fabre is by no means the misanthrope that some have chosen to think him. He delights in the society of women, and knows how to welcome them gracefully; and more than any one he is sensitive to the pleasant and stimulating impressions produced by the conversation of cultivated people.

He is no less fond of the arts, provided he finds in them a sincere interpretation of life. This is why the theatre, with its false values, its tinsel and affectation, has to him seemed a gross deformation of the reality, ever since the day when at

CHAPTER 15.

Ajaccio he attended a performance of "Norma," in which the moon was represented by a round transparent disc, lit from behind by a lantern hanging at the end of a string, whose oscillation revealed by turns first the luminary and then the transparency. This was enough to disgust him for ever with the theatre and the opera, whose motionless choruses, contrasting with the sometimes frantic movement of the music, left him with a memory of an insane and illogical performance.

Nevertheless, he adored music, of which he knew something, having learned it, as he learned his drawing, without a master; but he preferred the naive songs of the country, or the melody of a flute; to the most scholarly concert-music. (15/16.) In the intimacy of the modest chamber which serves as the family salon, with its few shabby and old-fashioned pieces of furniture, he plays on an indifferent harmonium little airs of his own composition, the subjects of which were at first suggested by his own poetry. Like Rollinat, Fabre rightly considers that music should complete, accentuate, and release that which poetry has perforce left incomplete or indefinite. This is why he makes the bise laugh and sing and roar; why he imitates the organ-tones of the wind in the pines, and seeks to reproduce some of the innumerable rhythms of nature; the frenzy of the lizard, the wriggling of the stickle-back, the jumping gait of the frog, the shrill hum of the mosquito, the complaint of the cricket, the moving of the Scarabaei, and the flight of the Libellulae.

Too busy by day to find time for much reading, it was at night that he would shut himself up. Retiring early to his little chamber, with bare walls and bare tile floor, and a window opening to the garden, he would lie on his low bed, with curtains of green serge, and would often read far into the night.

This philosopher, to whose books the philosophers of the future will resort for new theories and original ideas, refuses to have any commerce with other philosophers, disdaining their systems and preferring to go straight to the facts. Even when he took up Darwin's "Origin of Species" he did little more than open the book; so wearisome and uninteresting, he told me, did he find the reading of it. On the other hand, he is full of the ancient philosophers, and as he did not read them very extensively in his youth and middle age, he has returned to them finally with love and predilection for "these good old books." Unlike many thinkers of the day, he is persuaded that we cannot with impunity dispense with classic studies; and he rightly considers that science and the humanities are not rivals, but allies. Above all he has a particular affection for Virgil; one may say that he is steeped in his poetry; and he knows La Fontaine by heart. The style of the latter is curiously like his own, and Fabre owns himself as his disciple; certainly La Fontaine's is the most active influence which his work reveals. He has a profound acquaintance with Rabelais, who was always his "friend" and who constantly crops up in his conversation and his chance remarks.

After these his intellectual foster-parents have been Courrier, Toussenel, of whom he is passionately fond, and Rousseau, of whom he cares for little but his "Lettres sur la botanique," full of such fresh impressions, in which we feel not the literary man but the "craftsman"; he also cherishes Michelet; so full of intuition, although he never handled actual things and knew nothing of the practice of the sciences; not learned, but overflowing with love; his magic pen, his powers of evocation, and his deft brushwork delight Fabre, despite the poverty and insufficiency of his fundamental facts (15/17.); sometimes Michelet had been his inspiration. The two do really resemble one another; Michelet was no less fitted than Fabre to play the confidant to Nature, and his heart was of the same mettle.

Since I have spoken of his favourites, let me also speak of his dislikes; Racine, whom he cannot bear; Molière, whom he does not really like; Buffon, whom he frankly detests for his too fluent prose, his ostentatious style, and his vain rhetoric. The only naturalist whom he might really have delighted in, had he possessed his works and been able to read them at leisure, is Audubon, the enthusiastic painter of the birds of America. In him he felt the presence of a mind and a temper almost identical with his own.

CHAPTER 16.

TWILIGHT.

How he has laboured in this solitude! For he considers that he is still far from having completed his task. He feels more and more that he has scarcely done more than sketch the history of this singular and almost unknown world. "The more I go forward," he wrote to his brother in 1903, "the more clearly I see that I have struck my pick into an inexhaustible vein, well worthy of being exploited." (16/1.)

What studies he has undertaken, what observations he has carried out, "almost at the same time, the same moment!" His laboratory is crowded with these subjects of experiments. "As though I had a long future before me"--he was then just eighty years old--"I continue indefatigably my researches into the lives of these little creatures." (16/2.)

Work in solitude seems to him, more and more, the only life possible, and he cannot even imagine any other.

"The outer world scarcely tempts me at all; surrounded by my little family, it is enough for me to go into the woods from time to time, to listen to the fluting of the blackbirds. The very idea of the town disgusts me. Henceforth it would be impossible for me to live in the little cage of a citizen. Here I am, run wild, and I shall be so till the end." (16/3.)

For him work has become more than ever an organic function, the true corollary of life. "Away with repose! For him who would spend his life properly there is nothing like work--so long as the machine will operate."

Is this not the great law for all creatures so long as life lasts?

Why should the man who has made a fortune, who has neither children nor relations, and who may die tomorrow, continue to work for himself alone, to employ his days and his energies in useless labours which will profit neither himself nor his kind?

Ask of the Halictus, which, no longer capable of becoming a mother, makes herself guardian of a city, in order still to labour within the measure of her means.

Ask of the Osmia, the Megachile, the Anthidium, which "with no maternal aim, for the sole joy of labour, strive to expend their forces in the accomplishment of their vain tasks, until the forces of life fail."

Ask of the bee, which inaction leaves passive and melancholy so that she presently dies of weariness; of the Chalicodoma, so eager a worker that she will "let herself be crushed under the feet of the passer-by rather than abandon her task."

Ask it of all nature, which knows neither halt nor repose, and who, according to the profound saying of Goethe "has pronounced her malediction upon all that retards or suspends her progress."

Let us then labour, men and beasts, "so that we may sleep in peace; grubs and caterpillars in that torpor which prepares them for the transformation into moths and butterflies, and ourselves in the supreme slumber which dissolves life in order to renew it."

Let us work, in order to nourish within ourselves that divine intuition thanks to which we leave our original impress upon nature; let us work, in order to bring our humble contribution to the general harmony of things, by our painful and meritorious labour; in order that we may associate ourselves with God, share in His creation, and embellish and adorn the earth and fill it with wonders. (16/4.)

CHAPTER 16.

Forward then! always erect, even amid the tombs, to forget our griefs. Fabre finds no better consolation to offer his brother, who has lost almost in succession his wife and his eldest daughter:

"Do not take it ill if I have not condoled with you on the subject of your recent losses. Tried so often by the bitterness of domestic grief, I know too well the inanity of such consolations to offer the like to my friends. Time alone does a little cicatrize such wounds; and, let us add, work. Let us keep on our feet and at work as long as we are able. I know no better tonic." (16/5.)

And this exhortation to work, which recurs so often in the first letters of his youth, was to be the last word of the last volume which so splendidly terminates the incomparable series of his "Souvenirs": "Laboremus."

...

Age has killed neither his courage nor his energies, and he continues to work with the same zeal at nearly ninety years of age, and with as much eagerness as though he were destined to live for ever.

Although his physical forces are failing him, although his limbs falter, his brain remains intact, and is giving us its last

CHAPTER 16.

fruit in his studies on the Cabbage caterpillar and the Glow-worm, which mark a sudden rejuvenescence of thought on his part, and the commencement of a new cycle of studies, which promise to be of the greatest originality.

To him the animal world has always been full of dizzy surprises, and the insects led him "into a new and barely suspected region, which is ALMOST ABSURD." (16/6.)

The glow-worms, motionless on their twigs of thyme, light their lamps of an evening, in the cool of the beautiful summer nights. What do these fires signify? How explain the mystery of this phosphorescence? Why this slow combustion, "this species of respiration, more active than in the ordinary state"? and what is the oxidizable substance "which gives this white and gentle luminosity"? Is it a flame of love like that which lights the Agaric of the olive-tree "to celebrate its nuptials and the emission of its spores"? But what reason can the larva have for illuminating itself? Why is the egg, already enclosed in the secrecy of the ovaries, already luminous?

"The soft light of the Agaric has confounded our ideas of optics; it does not refract, it does not form an image when passed through a lens, it does not affect ordinary photographic plates." (16/7.)

But here are other miracles:

"Another fungus, the Clathrix, with no trace of phosphorescence, affects photographic plates almost as quickly as would a ray of sunlight. The Clathrix tenebrosa does what the Agaricus olearius has no power to do." (16/8.)

And if the beacon of the Glow-worm recalls the light of the Agaric, the Clathrix reminds us of another insect, the Greater Peacock moth.

In the obscurity of a dark chamber this splendid moth emits phantasmal radiations, perhaps intermittent and reserved for the season of nuptials, signals invisible to us, and perceptible only to those children of the night, who may have found this means to communicate one with another, to call one another in the darkness, and to speak with one another. (16/9.)

Such are the interesting subjects which only yesterday were occupying this great worker; the occult properties, the radiant energies of organic matter; of phosphorescence, of light, the living symbols of the great universal Eros.

But embarrassment long ago succeeded the ephemeral prosperity which marked the first years of his installation at

CHAPTER 16.

Sérignan, and that period of plenty was followed by a period of difficulty, almost of indigence. His class-books, which had succeeded marvellously, and from which the royalties had quickly attained to nearly 640 pounds sterling, which was the average figure for nearly ten years, were then no longer in vogue. Already the times had changed. France was in the crisis of the anti-clerical fever. Fabre made frequent allusions in his books of a spiritual nature, and many primary inspectors could not forgive what they regarded as a blemish.

We must also mention the keen competition caused by the appearance of similar books, usually counterfeit, and the more harmful for that; and as their adoption depended entirely on the caprice of commissions or the choice of interested persons, those of Fabre were gradually ceasing to sell.

It was from 1894 especially that their popularity declined so rapidly:

"Despite all my efforts here I am more anxious than ever about the future," he wrote to his publisher on the 27th of January, 1899; "two more of my books are about to disappear, a prelude to total shipwreck...I begin to despair." (16/10.)

CHAPTER 16.   320

He was not the man to have saved much money; numerous charges were always imposing themselves on him, and his first wife, careless of expenditure, had been somewhat extravagant.

While his position as teacher deteriorated his "Souvenirs" brought him little more than a nominal profit; for to most people he was still completely unknown among the potentates who monopolize the attention of the crowd.

"Work such as a Réaumur might be proud of will leave me a beggar, that goes without saying, but at least I shall have left my grain of sand. I would long ago have given up in despair, had I not, to give me courage, the continual research after truth in the little world whose historian I have become. I am hoarding ideas, and I make shift to live as I can." (16/11.)

Yet his reputation had long ago crossed the frontiers of his country. He had been a corresponding member of the Institute of France since 1887, and a Petit d'Ormoy prizeman. (16/12.) He was a member of the most celebrated foreign academies, and the entomological societies of the chief capitals of Europe; but his fame had not passed the walls of these academies and the narrow boundaries of the little world of professional biologists and philosophers.

CHAPTER 16.

Even in these circles, where he was almost exclusively read and appreciated, he was little known, and although he was much admired, although he was readily given credit for his admirable talent and exceptional knowledge, his readers were far from realizing the real powers of this world of life which he has called into being. His books are of those whose fertilizing virtues remain long hidden, to shine only at a distance, when much frothy writing, that has made a sudden noise in its time, has fallen into oblivion.

Every two or three years, after much fond polishing, he would open the door to yet another volume which was ready to go forth; adding astonishing chapters of the history of insects, wonderful fragments of animal psychology, but always obtaining only the same circumscribed success; that is, exciting no public curiosity, and remaining unperceived in the midst of general indifference.

His books interested only a select class, who, it is true, welcomed them eagerly, and read them with wonder and delight. If they excited the curiosity of a few philosophers, of scientists and inquirers, and here and there determined a vocation, still more, perhaps, did they charm writers and poets; they consoled Rostand at the end of a serious illness, their virtue, in some sort healing, procuring him both moral repose and a delightful relaxation. (16/13.) For all these, we may say, he has been one of those ten or twelve authors whom one would wish to take with one into a long exile, were they reduced to choosing no more before leaving civilization for ever.

Yet we must admit that this work has certain undeniable faults. The title, in the first place, has nothing alluring about it, and is calculated to deter rather than to attract purchasers, by evoking vague ideas of repulsive studies, too arduous or too special.

People have no idea of the wonderful fairyland concealed by this unpopular title; no conception that these records are intended, not merely for the scientist pure and simple, but in reality for every one.

Moreover, the first few volumes were in no way seductive. They boasted not the most elementary drawings to help the reader; not the slightest woodcut to give a direct idea of the insects described; of their shape, aspect, or physiognomy; and a simple sketch, however poor, is often worth more than long and laborious descriptions. The first volumes especially, printed economically, at the least possible expense, were not outwardly attractive.

It is also true that he had never founded any great hopes on the sale of such works.

CHAPTER 16.

Very few people are really interested in the lower animals, and Fabre has been reproached with wasting his time over "childish histories, unworthy of serious attention and unlikely to make money," of wasting in frivolous occupations the time which is passing so quickly and can never return. And why should he have still further wasted so many precious hours in executing minute drawings whose reproduction would have involved an expenditure which his publisher would not dare to venture upon, and which he himself could not afford?

For this universal inquirer was well fitted for such a task, and all these creatures which he had depicted he is capable of representing with brush and pencil as faithfully as with his pen. He had it in him to be not only a writer, but an excellent draughtsman, and even a great painter. He has reproduced in water-colour, with loving care, the decorations of the specimens of prehistoric pottery which his excavations have revealed, and which he has endeavoured to reconstruct, with all the science of an archaeologist. He has displayed the same skill in water-colour in that astonishing iconography, in which he has detailed, with marvellous accuracy, all the peculiarities of the mycological flora of the olive- growing districts. (16/14.)

As for those "paltry figures" insufficient or flagrantly incorrect in drawing, with which many people are satisfied, he regards them as "intolerable" in his own books, and as absolutely contradicting the rigorous accuracy of his text. (16/15.)

Of late years photography and the skill of his son Paul have supplied this deficiency. He taught his son to fix the insects on the sensitive plate in their true attitudes, in the reality of their most instantaneous gestures. However valuable such documents may be, how much we should prefer fine drawings, giving relief not only to forms and colours, but also to the most characteristic features and the whole living physiognomy of the creature! This is the function of art; but the great artist that was in Fabre was capable in this domain of rivalling the magical talent of an Audubon.

Such work was relinquished, although so many romances of nature, so much dishonest patch-work, won the applause due to success.

Fabre fell more and more into a state bordering on indigence, and finally he was quite forgotten. An opponent of evolution, he was out of the fashion. The encyclopaedias barely mentioned him. Lamarckians and Darwinians, who still made so much noise in the world, ignored him; and no

CHAPTER 16.

one came now to open the gate behind which was ageing, in obscurity and deserted, "one of the loftiest and purest geniuses which the civilized world at that moment possessed; one of the most learned naturalists and one of the most marvellous of poets in the modern and truly legitimate sense of the word." (16/16.)

In the department of Vaucluse, where he lived for more than sixty years, in Avignon itself, where he had taught for twenty years, the prefect Belleudy, who had succeeded in approaching him, was astonished and distressed to find "so great a mind so little known"; for even those about him scarcely knew his name. (16/17.)

But what matter! The hermit of Sérignan was not discouraged; he was disturbed only by the failure of his strength, and the fear that he could not much longer exercise that divine faculty which had always consoled him for all his sorrows and his disappointments. He could scarcely drag his weary limbs across the pebbles of his Harmas; but he bore his eighty-seven years with a fine disdain for age and its failings, and although the fire of his glance and that whole, eager countenance still expressed his passion for the truth, his abrupt gestures, touched with irony, his simple bearing, and the extreme modesty of his whole person, spoke sufficiently of his profound indifference toward outside contingencies, for the baubles of fame and all the stupidities of life.

At a few miles' distance, in another village, that other great peasant, Mistral, the singer of Provence, the poet of love and joy, the minstrel of rustic labour and antique faiths, was pursuing, amid the homage of his apotheosis, the incredible cycle of his splendid existence.

This glory had come to him suddenly; this fame "whose first glances are sweeter than the fires of dawn," and which was never to desert him for fifty long years.

The wind of favour which had sweetened his youth continued to propel him in full sail. He had only to show himself to be at once surrounded, felicitated, worshipped; and his mere presence would sway a crowd as the black peaks of the high cypresses are swayed by the great wind that bears his name. Like Fabre, he had remained faithful to his native soil; that soil which the great naturalist had never been able to leave without at once longing impatiently to return to its dusty olives where the cigale sings, its ilex trees and its thickets; and so he lived far from the cities, in a quiet village, with the same horizon of plains and hills that were balmy with thyme, leading in his little home an equal life full of wisdom and simplicity.

CHAPTER 16.

The hermit of Sérignan was the Lucretius of this Provence, which had already found its Virgil. With a very different vision, each had the same rustic tastes, the same love of the free spaces of wild nature and the scenes of rural life. But Mistral, wherever he looked, saw human life as happy and simple, through the prism of his creative imagination and the optimism of his happy life. Fabre, on the contrary, behind the sombre realities which he studied, saw only the ferocious engagement of confused living forces, and a frightful tragedy.

Thus their two lives, which were like parallel lines, never meeting, were in keeping with their work. And while Mistral, still young and triumphant despite the years, was at Maillane overwhelmed with honours and consideration, the poor great man of Sérignan lived an obscure and inglorious existence.

He had the greatest trouble to live and rear his family, and almost his sole income consisted of an uncertain sum of 120 pounds sterling annually, which he had for some years received, in the guise of a pension, by the generosity of the Institute, as the Gegner prize.

Finally his situation was so precarious that he decided to sell to a museum that magnificent collection of water-colour plates in which he had represented, life-size and with an astonishing truth of colour, all the fungi which grow in Provence.

He wrote to Mistral on the subject, after the visit which the latter paid him in the spring of 1908: the only visit of the kind. Before meeting in Saint-Estelle, the Paradise of the Félibres, they had wished not to die before at least meeting on this earth.

Fabre wrote to mistral the following letter, which I owe to the kindness of the great poet:--

"I have never thought of profiting by my humble fungoid water- colours...Fate will perhaps decide otherwise.

"In this connection, permit me to make a confession, to which your nobility of character encourages me. Until latterly I had lived modestly on the product of my school-books. To-day the weathercock has turned to another quarter, and my books no longer sell. So here I am, more than ever in the grip of that terrible problem of daily bread. If you think, then, that with your help and that of your friends, my poor pictures might help me a little, I have decided to let them go, but not without bitterness. It is like tearing off a piece of my skin, and I still hold to this old skin, shabby as it may be; a little for my own sake, much more for my family's, and much more again for the sake of my entomological studies, studies which I feel obliged to pursue, persuaded that for a long time to come no one will care to resume them, so ungrateful is the calling." (16/18.)

At the instigation of the poet the prefect Belleudy took it upon him to intercede with the Minister, from whom he finally wrung a grant of 40 pounds sterling, "in encouragement of the sciences." Finally he ventured to reveal the situation to the General Council of Vaucluse, and to require it to contribute at least its share, in order to ensure a peaceful and decent old age to a man who was not only the greatest celebrity of the department, but also one of the highest glories of the nation. He pleaded so well and so nobly that the assembly granted Fabre an annual sum of 20 pounds sterling, "as the public homage which his compatriots pay to his lofty science and HIS EXCESSIVE MODESTY." (16/19.) At the same time, in a generous impulse, the Council placed at his disposal all the scientific equipment of the departmental laboratory of agricultural analysis, which was no longer used; there was indeed talk of suppressing it.

Now that the burden of his days weighed so heavily on him, and his task was virtually finished, everything, by the customary irony of things, was coming his way simultaneously: not only what was necessary and indispensable, but even something that was superfluous.

CHAPTER 16.

So one day all these delicate instruments, useless to a biologist who by the very nature of his labours had done without them all his life, and had never wearied of denying their utility, arrived at Sérignan. He did not possess even one modest thermometer; and as for the superb microscope over which he so often bent, the only costly instrument in his rustic laboratory, it was a precious present which, at the instigation of Duruy, Dumas the chemist had given him years before; but a simple lens very often sufficed him. "The secrets of life," he somewhere writes, "are to be obtained by simple, makeshift, inexpensive means. What did the best results of my inquiry into instinct cost me? Only time, and above all, patience."

It was then that a few of his disciples, finally affected by such abandonment, decided to celebrate his jubilee, hoping thus to reveal both his name and his wonderful books to the crowd that knew nothing of him. (16/20.)

It was time; a little longer, and, according to his racy phrase, "the violins would have come too late." The old master is daily nearer his decline; his sight, once so piercing, is now so obscured that he can barely see to sign his name, in a small, tremulous hand, confused and illegible. His muscles are so feeble now that he can walk only in short steps, on his wife's arm, leaning on a cane; and he would soon be piteously exhausted were not some seat available within immediate reach. Very soon now he will no longer hope to make the tour of this Harmas, which his feet have trodden daily for thirty years. In this failure of the body, all that survives are the two sparkling cavities of his eyes and his extraordinary memory.

But he is far from being mournful: he feels only an immense lassitude, and an infinite regret that perhaps he will not be able to bring his series of "Souvenirs" to the point he had desired; not wishing to die until he has pushed his career as far as is in his power; without having worked, on his feet, until the very hour when the light of this world is suddenly withdrawn, and his eyes open upon the infinite life, beyond the infinite worlds of space.

The festival took place on the 3rd of April of the year 1910, and was touching in its simplicity.

What an unforgettable day in the life of Fabre! That morning the gate of the Harmas was left open to all, and many of the people of Sérignan who invaded the garden were able to look for the first time on the face of their fellow-citizen, who had so long lived among them, and whom they had now, to their astonishment, discovered.

But among the crowd of friends and admirers who, coming from all parts, pressed around the little pink house, the most amazed of all was Marius, the blind cabinet-maker, unable to contain his intense delight at the sudden burning of so much incense before his idol, for to him it had seemed that this day of apotheosis would never dawn!

For nothing was certain, although the day of the jubilee had long been fixed. In the first place there had been serious defections in the ranks of the official personages who were to take part in the ceremony. Then the weather was terrible for the time of year; the spring had commenced gloomily, a season of floods and catastrophes. But on this morning the rain of days had ceased to fall, and suddenly the sun appeared.

Among other compliments and marks of homage the old man was presented with a golden plaque, on one side of which Sicard, who stood revealed as a master of the burin, had engraved his portrait with rare fidelity. The reverse was resplendent with one of the most beautiful syntheses which the history of art has known; a surprising allegory, in which the imagination of the artist evoked the man of science, the singer of the insects, the landscape which had seen the birth of so many little lives, and the village amid the olive-trees, in front of the sun-steeped Ventoux.

At this festival, the jubilee of a scientist, the scientists were least numerous.

CHAPTER 16.

The banquet was given in the large room of a cafe in the midst of Sérignan; in order, no doubt, that in this humble life even glory should be modest.

As Fabre could not walk, he was helped into the carriage of ceremony, which was sent expressly from Orange, and the little procession, which was swelled by the municipal choral society, spurred on by Marius, moved slowly off along the sole central street.

It was a great family repast: one of those love-feasts in which all communicate in a single thought.

Edmond Perrier brought the naturalist the homage of the Institute, and expressed in unaffected terms the just admiration which he himself felt. The better to praise him, he gave a summary of his admirable career, and his immortal work. At the evocation of this long past of labour Fabre regretted his poor vanished joys, "the sole moments of happiness in his life."

Moved to tears, by his memories and by the simple and pious homage at last rendered to his genius, he wept, and many, seeing him weep, wept with him.

Others spoke in the name of the great anonymous crowd of friends, of all those who had found a source of infinite enjoyment in his works. At the same time the greatest writers, the greatest poets sent on the same day, at the same hour, their salutation or eloquent messages to the "Virgil of the insects" (16/21.), to the "good magician who knew the language of the myriad little creatures of the fields." (16/22.)

Doubtless he would sooner or later have received full justice; but without this circumstance it is permissible to add that the end of his life would have passed amidst the completest oblivion, and that he would have taken leave of the world without attracting any particular attention. His death would have occurred unperceived, and when the little vault of Vaison stone, up in the small square enclosure of pebbles which serves as the village cemetery, where those he has loved await him, came to be opened for the last time, they would hardly have troubled to close it again.

Yet the honours paid him were far from being such as he merited.

Why, at this jubilee of the greatest of the entomologists, was not a single appointed representative of entomology present? (16/22.)

The fact is that the majority of those who "amid the living seek only for corpses," according to the expression of Bacon, unwilling to see in Fabre anything more than an imaginative writer, and being themselves incapable of understanding the beautiful and of distinguishing it in the true, reproached him, perhaps with more jealousy than conviction, with having introduced literature into the domains of science.

Other entomological specialists accuse him of presenting in the guise of science discoveries which have been made by others. But in the first place, as he has read very little, he certainly did not know all that had been done by others; and what matter if he had discovered nothing essential concerning this or that insect if the result of his study of it has been to impregnate it with something new, or to touch it with the breath of life?

Others, finally, who wished to see with their own eyes the proof of his statements, have reproached him with a few errors; but he observed so skilfully that these errors, if any have really slipped into his books, cannot be very serious.

He was one of the glories of the University, but it failed to add to the brilliance of this ceremony, and it is to be regretted that the Government could not amid its temporary preoccupations have done with all the spontaneity that might have been looked for the one thing which might on this memorable date have atoned for its unjust obliviousness. Since Duruy had created Fabre a chevalier of the Empire more than forty years had gone by, and in this long interval Fabre was absolutely ignored by the authorities. While the State daily raises so many commonplace men to the highest honours, it was afterwards needful to procure the intervention of influential persons, to justify his worth and to prove his deserts, in order to obtain his promotion through one degree of rank in that Legion of Honour which his eminent services had so long adorned.

This tardy reparation at least had the result of shedding a twilight of glory over the evening of his life, and from that day he suddenly appeared in his true place and took his rank as a man of the first order. Everybody began to read him, and presently no one was willing to seem ignorant of him, for more of his "Souvenirs entomologiques" were sold in a few months than had been disposed of in more than twenty years. (16/24.)

At last Fabre experienced not only glory and renown, but also popularity. This was only justice, for his is essentially a popular genius. Has he not striven all his life to place the marvels of science within reach of all? And has he not written above all for the children of the people?

CHAPTER 16.

So at last people have learned the way to the Harmas; they go thither now in crowds, to visit the enclosure and the modest laboratory, as to a veritable place of pilgrimage which attracts from afar many fervent admirers.

Some, it is true, go thither to see him simply as an object of curiosity; but even among these there are those who on returning thence, full of enthusiasm for what they have seen, find the flowers of the fields more sweet and fragile, and the wild fragrance of the woods and hedges more voluptuous, and the green of the trees more tender. They have learnt to look at the earth and to "kneel in the grass."

Scientists come to chat with the scientist. Others come to salute the primary schoolman, the lay instructor, the great pedagogue whose glory is reflected upon all the primary schools of France.

Those who cannot visit him write, telling him of all the pleasure which they owe him, thanking him for long and delightful hours passed in the reading of his books, expressing the hope that he may yet live many years, and still further increase the number of his "Souvenirs."

Some ask him a host of questions relating to entomology or philosophy; others ask him for impossible answers to some of the fascinating and mysterious problems which he has expounded; women confide in him their little private griefs or their intimate sorrows, a naive form of homage; but a thousand times more touching than any other, and one that shows how profound has been the beneficent influence of his books upon certain isolated minds, and what consolation can be derived from science when it finds a sufficiently eloquent voice to interpret it.

As he can work no longer, these visits now fill his life, formally so occupied; and in the midst of all the sympathy extended to him he is sensible, not of the twilight, but of a sunrise; he feels that his work has been good, that an infinity of minds are learning through him to regard plants and animals with greater affection; and that the consideration of men, finally directed upon his work, will not readily exhaust it, for it is one of the Bibles of Nature.

NOTES.

NOTES TO INTRODUCTION.

Introduction/1. Letters to his brother, 1898-1900.

Introduction/2. I have made some valuable "finds" here; among other pieces cited the fragment on "Playthings," the curious description of the "Eclipse," and the poem on "Number" are here published for the first time.

Introduction/3. This negligence in the matter of correspondence is not least among the causes which have mitigated against his popularity.

NOTES TO CHAPTER 1.

1/1. "It is a country that has very little charm." To his brother, 18th August, 1846.

1/2. "Practicien, homme d'affaires ou de chicane": roughly, "practitioner, man of business or law": so his father is described in his birth certificate.

1/3. "Souvenirs entomologiques," 2nd series, chapter 4, and 7th series, chapter 19.

1/4. Id., 8th series, chapter 8.

1/5. To his brother, 15th August, 1896.

1/6. Id. "As brothers, we are one only; but in virtue of our different tastes we are two, and I am amused and interested where you might well be bored."

1/7. Frédéric Fabre, like his brother, an ex-scholar of the normal primary school of Vaucluse, was first of all teacher at Lapalud (Vaucluse), then professor in the communal college of Orange. He was director of the primary school attached to the normal school of Avignon, where he voluntarily retired from teaching in 1859. He then became, successively, secretary to the Chamber of Commerce of Avignon, director of the Vaucluse Docks, and finally director of the Crillon Canal, which position he still occupies (December, 1912).

1/8. "Souvenirs entomologiques," 10th series, chapter 9.

1/9. Among his innumerable manuscripts I have found a vast number of little poems, which date from this period.

1/10. It was then that he gave up his position to his brother Frédéric, who had continually followed closely in his steps, and who in turn had just obtained the qualification of pupil-teacher and bursar (August, 1842).

1/11. "Souvenirs entomologiques," 10 series, chapter 21.

1/12. To his brother, 2nd and 9th of June, 1851.

NOTES TO CHAPTER 2.

2/1. "Souvenirs entomologiques," 1st series, chapter 20, and 9th series, chapter 13.

CHAPTER 16. 341

2/2. Id., 6th series, chapter 21.

2/3. To his brother, from Ajaccio, 10th June, 1850.

2/4. Id., id.

2/5. Id., from Carpentras, 15th August, 1846.

2/6. Id., from Ajaccio, 10th June, 1850.

2/7. Id., from Carpentras, 15th August, 1846.

2/8. Id., id.

2/9. "Souvenirs entomologiques," 1st series, chapter 14.

2/10. To his brother, from Carpentras, 3rd September, 1848.

2/11. Id., 8th September, 1848.

2/12. Id., id.

2/13. Id., 3rd September, 1848.

2/14. Id., id.

2/15. Letter to the Rector of the Nîmes Academy, 29th September, 1848.

2/16. To his brother, 29th September, 1848.

NOTES TO CHAPTER 3.

3/1. To his father, from Ajaccio, 14th April, 1850.

3/2. To his brother, from Ajaccio, 1851.

3/3. To his brother, from Ajaccio, 9th June, 1851. "I have set to work upon a conchology of Corsica, which I hope soon to publish."

3/4. The Helix Raspaillii.

3/5. To his brother, from Ajaccio, 10th June, 1850.

3/6. Id., id.

3/7. "Souvenirs entomologiques," 9th series, chapter 14.

3/8. Number, (Le Nombre--ARITHMOS), poem, Ajaccio, September, 1852.

3/9. To his brother, from Ajaccio, 2nd June, 1851.

CHAPTER 16. 343

3/10. Id., 10th October, 1852, and "Souvenirs entomologiques," 10th series, chapter 21.

3/11. Fr. Mistral, "Mémoires." Moquin-Tandon, born at Montpellier, was professor of Natural History at Marseilles, at Toulouse, and in Paris.

3/12. To his brother, from Ajaccio, 10th October, 1852.

3/13. Id.

3/14. To his brother, from Carpentras, 3rd December, 1851. "Our crossing was atrocious. Never have I seen so terrible a sea, and that the packet-boat was not broken up by the force of the waves must have been due to the fact that our time had not yet come. On two or three occasions I thought my last moment was at hand; I leave you to imagine what a terrible experience I had. In ordinary weather the packet by which we travelled makes the voyage from Ajaccio to Marseilles in about eighteen hours; it is said to be the fastest steamer on the Mediterranean. On this occasion it took three days and two nights."

3/15. January, 1853.

NOTES TO CHAPTER 4.

4/1. To his brother, from Avignon, 1st August, 1854. "I have arrived at Toulouse, where I have passed the best examination one could possibly wish. I have been accepted as licentiate with the most flattering compliments, and the expenses of the examination should be returned to me. The examination was of a higher level than I had expected."

4/2. To M. -- (of the Institute), from Avignon, 1854. (Letter communicated to M. Belleudy, prefect of Vaucluse, by M. Vollon, painter.)

4/3. Id.

4/4. To his brother, from Ajaccio, 10th October, 1852.

4/5. Observations concerning the habits of the Cerceris and the cause of the long preservation of the coleoptera with which it provisions its larvae.--"Annales de Sc. natur.," 4th series, 1855.

4/6. "Souvenirs entomologiques," 10th series, chapter 22.

4/7. "I had only one idea: to free myself, to leave the lycée, where, not being a fellow, I was treated as a subordinate. An inspector-general told me frankly one day, 'You will never amount to anything if you are not a fellow' (agrégé).

CHAPTER 16. 345

'These distinctions disgust me,' I replied." (Conversations.)

4/8. To his brother, from Ajaccio, 14th January, 1850.

4/9. Inquiries respecting the tubercles of Himantoglossum hircinum. Thesis in Botany, 1855.

4/10. Inquiries respecting the anatomy of the reproductive organs, and the developments of the Myriapoda. Thesis in Zoology, 1855.

4/11. Prize for experimental physiology, 1856.

4/12. Letter to Léon Dufour, 1st February, 1857.

4/13. "The Origin of Species," 1857 (?), translated by Barbier, page 15.

4/14. "Souvenirs entomologiques," 1st series, chapter 1, and 5th series, chapter 1.

4/15. Id., 1st series, chapter 16.

4/16. Id., 1st series, chapter one.

4/17. Henry Devillario, magistrate at Carpentras, where he performed his duties as juge d'instruction until his death. A

CHAPTER 16.   346

notable collector and distinguished publicist. Dr. Bordone, to-day at Frontignan. Vayssières, professor of Zoology in the faculty of sciences at Marseilles.

4/18. "Souvenirs entomologiques," 1st series, chapter 13.

4/19. He was subject in his youth to violent headaches, "which sometimes developed into a cerebral fever," as well as strange nervous troubles: "A few days ago I was attacked, at night, with a sudden nervous illness, of a terrifying nature, which I have not as yet been able to identify." To his brother, 3rd September, 1848. Severe disappointment or annoyance always had a great effect upon him; on the occasion of his first marriage he fell into a sort of cataleptic condition as a result of the opposition of his parents and relations, who sought to oppose it. (Conversations with his brother.)

4/20. "Souvenirs entomologiques" 9th series, chapter 23.

4/21. Id., 10th series, chapter 22.

4/22. Letter to Lèon Dufour, 1st February, 1857. "Steps have been taken to obtain for me the post of drawing-master (maître des travaux graphiques). If they succeed, thanks to the little talent I have for drawing, my salary will reach a reasonable figure, 120 pounds sterling, and I can then, by giving up these abominable private lessons, cultivate rather more seriously the studies into which you have initiated me." Communicated by M. Achard.

4/23. "Souvenirs entomologiques" 10th series, chapter 22.

4/24. Oubreto Prouvençalo. La Cigale et la Fourmi.

4/25. Lavisse. A minister. Victor Duruy.

4/26. Letter to the municipal councillors of Avignon.

4/27. J. Stuart Mill, "Autobiography," chapter 6.

4/28. I have visited this house; nothing, at all events outside, has changed in the least.

4/29. Mill collaborated in his "Flore du Vaucluse": "A virtuous man whose recent loss we shall all deplore joined his efforts to mine in this undertaking." Letter to the Mayor of Avignon, 1st December, 1833, communicated by M. Félix Achard.

NOTES TO CHAPTER 5.

5/1. "Chimie agricole."

CHAPTER 16. 348

5/2. "Le Ciel." Lectures et Leçons pour tous.

5/3. "La Terre." Lectures et Leçons pour tous.

5/4. "La Chimie de l'oncle Paul." Lectures courantes pour toutes les écoles.

5/5. "Histoire de la bûche."

5/6. "Les jouets. Le Toton" (manuscript). The primitive fountain, the "antique appliance" transmitted by inheritance, "the invention perhaps of some little unemployed herd-boy," consisted originally of three apertures and three straws; two similar apertures on one side, with two short straws, which dipped into the water, and a single orifice on the other side for the longer straw which delivered the water. Happening one day to use only two straws, one on each side, the little Fabre perceived that the device worked just as well, and "so, quite unconsciously, without thinking of it, I discovered the syphon, the true syphon of the physicist." Loco cit.

5/7. "The chemistry course is a great success at home." To his brother, from Orange, 1875.

5/8. To his son Émile, 4th November, 1879. "The household; discussions as to domestic economy for use in girls' schools."

5/9. "Souvenirs entomologiques," 2nd series, chapter 1.

5/10. To the Mayor of Avignon, 1st December, 1873. Communicated by M. Félix Achard.

5/11. Letter to his brother, 1875.

5/12. Id.

NOTES TO CHAPTER 6.

6/1. "Souvenirs entomologiques," 2nd series, chapter 1. "L'Harmas."

6/2. Id., 6th series, chapter 5.

6/3. The Lumbricus phosporeus of Dugés. Fabre had already clearly perceived that this curious phenomenon of phosphorescence appears at birth, and he saw in it a process of oxidation, a species of respiration, especially active in certain tissues. Letter to Léon Dufour, 1st February, 1857. Communicated by M. Félix Achard.

6/4. To his brother, from Carpentras, 15th August, 1846.

CHAPTER 16. 350

6/5. He died at the age of 96.

6/6. "Souvenirs entomologiques," 1st series, chapter 21.

6/7. To his son Émile, 4th November, 1879.

6/8. To Henry Devillario, 30th March, 1883.

6/9. Id., 17th December, 1888.

NOTES TO CHAPTER 7.

7/1. "Souvenirs entomologiques," 8th series, chapter 12.

7/2. Id., 7th series, chapter 16.

7/3. Id., 1st series, chapter 4.

7/4. Id., 2nd series, chapter 3.

7/5. Id., 6th series, chapter 21.

7/6. Id., 1st series, chapter 19, and 2nd series, chapter 7.

7/7. Id., 7th series, chapter 23.

7/8. Maeterlinck, "The Bee."

7/9. "Souvenirs entomologiques," 7th series, chapter 2.

7/10. Id., 8th series, chapter 22.

7/11. Id., 6th series, chapter 6.

7/12. Id., 9th series, chapter 10.

7/13. Bergson, "l'Evolution créatrice."

7/14. "Souvenirs entomologiques," 10th series, chapter 6.

7/15. "Les Serviteurs" and "Les Auxiliaires."

7/16. François Raspail, born at Carpentras in 1794, was also a professor at the college of Carpentras.

7/17. To his brother, 3rd September, 1848. The improvement did not last long; the child died finally a short time afterwards.

7/18. "Souvenirs entomologiques," 10th series, chapter 21.

7/19. Ed. Perrier. Private letter, 27th October, 1909. "He is the finest of all our observers, and all scientists should bow to the facts which he excels in discovering."

7/20. "Souvenirs entomologiques," 6th series, chapter 25.

7/21. Id., 10th series, chapter 16.

7/22. Id., 10th series, chapter 20.

7/23. Manuscripts, unpublished observations.

7/24. A common spectacle in Provence, but one which Fabre never wearied of seeing.

7/25. "Souvenirs entomologiques," 6th series, chapter 17.

7/26. We know that the great naturalist was far from being charmed by the song of the nightingale.

7/27. Manuscripts, unpublished observation. These remarks deal with the solar eclipse of 28th May, 1900.

7/28. Among the insects which he has observed there are many which are not always sufficiently characterized. "Insectes coléoptères observes aux environs d'Avignon." Avignon, pub. Seguin, 1870.

7/29. Coleoptera observed in the neighbourhood of Avignon. A catalogue now very scarce, a copy of which I owe to the kindness of Dr. Chobaut, of Avignon.

7/30. Nomina si nescis, perit et cognitio rerum.

7/31. "Souvenirs entomologiques," 4th series, chapter 11.

7/32. Id., 9th series, chapter 19.

7/33. Id., 1st series, chapter 9.

7/34. "Jenner's Legend of the isolation of the young Cuckoo in the nest," by Xavier Raspail, "Bull. de la Soc. Zool. de France," 1903.

7/35. "Souvenirs entomologiques" 1st series, passim.

7/36. Id., 4th series, chapter 14.

7/37. Id., 1st series, chapter 7.

7/38. Id., 2nd series, chapter 2.

NOTES TO CHAPTER 8.

8/1. "Souvenirs entomologiques" 1st series, chapter 2.

8/2. Bergson, "l'Evolution créatrice."

8/3. "Souvenirs entomologiques," 2nd series, chapter 4.

CHAPTER 16.   354

8/4. Id., 5th series, chapter 8.

8/5. Id., 9th series, chapter 3.

8/6. Id., 1st series, chapter 22.

8/7. Id., 4th series, chapter 3.

8/8. Id., 4th series, chapter 3.

8/9. Id., 4th and 1st series, chapter 19.

8/10. Id., 9th series, chapter 24.

8/11. Id., 10th series, chapter 5.

8/12. Id., 4th series, chapter 6.

8/13. Id., 9th series, chapter 16.

8/14. Id., 2nd series, chapter 5.

8/15. Id., 5th series, chapter 7.

8/16. Id., 6th series, chapter 8.

8/17. Id., 3rd series, chapters 17, 18, 19 and 20.

8/18. Id., 2nd series, chapter 15.

8/19. Id., 3rd series, chapter 11.

8/20. Emerson.

8/21. "Souvenirs entomologiques," 4th series, chapter 9.

8/22. Unpublished observations.

8/23. "Mireille," 3rd canto.

NOTES TO CHAPTER 9.

9/1. "Souvenirs entomologiques," 8th series, chapter 21.

9/2. "Les Ravageurs," chapter 34, agriculture.

9/3. "Souvenirs entomologiques," 10th series, chapter 12.

9/4. Id., 1st series, chapter 2, and 10th series, chapter 13.

9/5. Id., 2nd series, chapter 17.

9/6. Id., 7th series, chapter 20.

9/7. Id., 2nd series, chapter 4.

CHAPTER 16. 356

9/8. At novitas mundi nec frigora dura ciebat, Nec nimios aestus. Lucretius, "De Natura rerum."

9/9. In this connection see the excellent introduction written by M. Edmond Perrier to serve as preface to the work of M. de Romanes: "l'Intelligence des animaux."

9/10. "Souvenirs entomologiques," 8th series, chapter 20.

9/11. To Henry Devillario, 30th March, 1883.

9/12. To Henry Devillario, 12th May, 1883.

9/13. To his brother, 1900.

9/14. Letters to his brother. "I am not sulking; far from it...I have no lack of ink and paper; I am too careful of them to lack them; but I do lack time...So you still think I am sulking because I do not reply! But imagine, my dear and petulant brother, that for several weeks I have been pursuing, with unequalled persistence, some abominable conic problems proposed at the fellowship examination, and once I have mounted my hobby-horse, good-bye to letters, good-bye to replies, goodbye to everything." (Carpentras, 27th November, 1848.) "You are right, seven times right to storm at me, to grumble at my silence, and I admit, in all contrition, that I am the worst correspondent you could find.

To force myself to write a letter is to place myself on the rack, as well you know...But why do you get it into your head, why do you tell me, that I disdain you, that I forget you, that I ignore you, you, my best friend?...For my silence blame only the multiplicity of tasks, which often surpasses, not my courage, but my strength and my time." (Ajaccio, 1st June, 1851.)

9/15. "Souvenirs entomologiques," 10th series, chapter 8.

9/16. Id., 9th series, chapter 2.

NOTES TO CHAPTER 10.

10/1. "Souvenirs entomologiques," 1st series, chapter 21.

10/2. Id., 9th series, chapter 2.

10/3. Id., 10th series, chapter 4.

10/4. Montaigne's Essays.

10/5. "Souvenirs entomologiques," 8th series, chapter 17.

10/6. "Les Ravageurs."

10/7. "Souvenirs entomologiques," 10th series, chapter 18, and "Merveilles de l'instinct: la Chenille du chou."

10/8. Id., 8th series, chapter 17.

NOTES TO CHAPTER 11.

11/1. "Souvenirs entomologiques," 3rd series, chapter 8.

11/2. Id., 2nd series, chapter 14 et seq.

11/3. Id., 6th series, chapter 9.

11/4. Id., 5th series, chapter 19.

11/5. Tolstoy: "All that the human heart contains of evil should disappear at the contact of nature, that most immediate expression of the beautiful and the good." ("The Invaders.")

11/6. The "Livre d'histoires" and "Chimie agricole."

11/7. "Oubreto Provençalo. La Bise."

11/8. Id., "Le Semeur."

11/9. Id., "Le Crapaud."

NOTES TO CHAPTER 12.

12/1. "Oubreto Provençalo. Le Maréchal."

12/2. "Oubreto Provençalo."

12/3. In this connection see the admirable passage in Sainte-Beuve's "Port- Royal," Book 2, chapter 14.

12/4. "Souvenirs entomologiques," 4th series, chapter 1.

12/5. Id., 1st series, chapter 17.

12/6. Id., 7th series, chapter 8.

12/7. Id., 7th series, chapter 10.

12/8. Id., 8th series, chapter 8.

12/9. Id., 8th series, chapter 20.

12/10. Id., 6th series, chapter 14.

12/11. Id., 8th series, chapter 18.

12/12. Id., 10th series, chapter 8.

12/13. Id., 10th series, chapter 6.

12/14. Id., 5th series, chapter 22.

NOTES TO CHAPTER 13.

13/1. "Souvenirs entomologiques," 10th series, chapter 17.

13/2. Id., 9th series, chapter 4, "l'Exode des arignées" (the Exodus of the Spiders), and chapter 5, "l'Araignée crabe" (the Crab Spider).

13/3. Id., 5th series, chapter 17.

13/4. Id., 3rd series, chapter 8.

13/5. Id., 6th series, chapter 14. "Oubreto. Le Grillon," and unpublished verses.

13/6. "Souvenirs entomologiques," 2nd series, chapter 16.

13/7. Id., 9th series, chapter 21.

13/8. "Les Merveilles de l'instinct: le Ver luisant" (Marvels of Instinct: the Glow-worm).

13/9. "Souvenirs entomologiques," 2nd series, chapter 12.

13/10. Id., 8th series, chapter 22, and 9th series, chapter 11.

13/11. Id., 5th series, chapter 18.

NOTES TO CHAPTER 14.

14/1. Grandjean de Fouchy: eulogy of Réaumur, in "Recueils de l'Acad.des sciences," volume 157 H, page 201, and Preface to the "Lettres inédites de Réaumur," by G. Musset.

14/2. "Mémoires," passim, and volume 2, 1st mémoire.

14/3. Id., volume 3, 3rd mémoire.

14/4. Id., volume 2, 1st mémoire. Ch. Tellier, "Le Frigorifique" (Refrigeration), story of a modern invention, chapter 23; cold applied to the animal kingdom.

14/5. Léon Dufour: "Journal de sa vie." Souvenirs and impressions of travel in the Pyrenees to Gavarnie, Héas, the "Montagnes maudites," etc. Entomological excursions on the dunes of Biscarosse and Arcachon.

14/6. Id., direction of entomological studies.

14/7. "Souvenirs entomologiques" 2nd series, chapter 1: "L'Harmas."

14/8. Id., 5th series, chapter 11.

NOTES TO CHAPTER 15.

15/1. Louis Charrasse, private letter, 20th February, 1912, and "Le Bassin du Rhône," March, 1911.

15/2. "Oubreto. Le Crapaud."

15/3. It was only in the afternoon that he devoted himself, when needful, to microscopic researches, on account of the better inclination of the light.

15/4. He lost it at the end of last spring.

15/5. "Les Serviteurs. Le Canard."

15/6. "Souvenirs entomologiques," 1st series, chapter 13: an ascent of Mont Ventoux.

15/7. The name given to Christmas in Provence.

15/8. Louis Charrasse, private letters.

CHAPTER 16.

15/9. Id.

15/10. 1888-1892.

15/11. "Souvenirs entomologiques," 2nd series, chapter 2.

15/12. Louis Charrasse, private letter.

15/13. Letter to his nephew, Antonin Fabre, 4th January, 1885.

15/14. "Souvenirs entomologiques," 6th series, chapter 19.

15/15. Id., 6th series, chapter 2.

15/16. Id., 6th series, chapter 11.

15/17. Conversations.

NOTES TO CHAPTER 16.

16/1. Letter to his brother, 4th February, 1900.

16/2. To his brother, 18th July, 1908. At this time the eighth volume of his "Souvenirs" had just appeared, and the ninth was in hand.

CHAPTER 16. 364

16/3. Id.

16/4. "Chimie agricole."

16/5. To his brother, 10th October, 1898.

16/6. Private letter, 30th March, 1908.

16/7. Id.

16/8. Id.

16/9. Unpublished experiments.

16/10. To Charles Delagrave, 27th January, 1899.

16/11. To his brother, 4th February, 1900.

16/12. This prize was awarded to Fabre in 1899. The amount of the prize is 400 pounds sterling. It is one of the chief prizes of the Institute.

16/13. Edmond Rostand. Private letter, 7th April, 1910: "His books have been my delight during a very long convalescence."

16/14. This magnificent atlas, the gem of Fabre's collections, comprises nearly 700 plates, and a large body of explanatory and descriptive matter.

16/15. To Charles Delagrave, undated.

16/16. Maeterlinck. Private letter, 17th November, 1909. "Les 4 Chemins, "Grasse (Alpes-Maritimes). "You overwhelm me with pleasure and do me the greatest honour in allowing my name to be inscribed among those of the committee which proposes to celebrate the jubilee of Henri Fabre...Henri Fabre is, indeed, one of the chiefest and purest glories that the civilized world at present possesses; one of the most learned naturalists and the most wonderful of poets in the modern and truly legitimate sense of the word. I cannot tell you how delighted I am by the chance you offer me of expressing in this way one of the profoundest admirations of my life."

16/17. J. Belleudy, prefect of Vaucluse. Private letter, 29th September, 1909. "It pains me to see so great a mind, so eminent a scientist, such a master of French literature, so little known. Two years ago, when the Gegner prize was awarded to him, I felt that I must speak of him to certain of those about me; and they had hardly heard his name!"

16/18. Letter to Frédéric Mistral, 4th July, 1908.

16/19. Council General of Vaucluse, session of August, 1908. The words of the recorder, M. Lacour, mayor of Orange, to-day deputy for Vaucluse, a personal friend and ardent admirer of the old master.

16/20. Edmond Rostand. Private letter, 20th November, 1909. "I am, sir, not only greatly touched, but also and above all delighted that you have thought of including me among the friends who wish to fete Henri Fabre. Thanks for having considered that my name would assist your undertaking. The "Souvenirs entomologiques" have long ago made me intimate with his charming, profound, and moving genius. I owe them an infinity of delightful hours. Perhaps also I ought to thank them for having encouraged one of my sons to pursue the vocation which he entered. If, in order to honour Henri Fabre, you run the pious risk of disturbing, for a moment, the studious retreat in which, for so many years, he has pursued his life and his work, it is an act of justice toward this great scientist, who thinks as a philosopher, sees as an artist, and feels and expresses himself as a poet." Romain Rolland. Private letter, 7th January, 1910. "You cannot imagine what pleasure you have given me by requesting me to associate myself in the glorification of J.H. Fabre. He is one of the Frenchmen whom I most admire. The impassioned patience of his ingenious observations delights me as much as the masterpieces of art. For years I have read and loved his books. During my last holidays, of three volumes that I travelled with two were volumes of his "Souvenirs entomologiques." You will honour me and delight me by counting me as one of you."

16/21. Edmond Rostand. Telegram.

16/22. Romain Rolland.

INDEX.

Achard, M.

Agaricus, luminosity of.

"Agricultural Chemistry."

Ajaccio, Fabre at.

Ammophila.

Anthidium.

Anthophora.

Anthrax.

CHAPTER 16. 368

Arachne clotho.

Arachnoids, cannibalism of.

Audubon.

Avignon, Fabre at. suggested agronomic station at.

Balaninus.

Balzac.

Bees.

Belleudy, M.

Bembex.

Bergson.

Bernard, Claude.

Blanchard.

Blue fly.

Bombyx.

Bordone.

Bossuet.

Bourdon.

Buffon.

Buprestis.

Calendal.

Calendar-beetle.

Calosoma sycophanta.

Candolle, de.

Cannibalism.

Cantharides.

Cantharis, courtship of.

Capricornis.

Carabidae.

CHAPTER 16.

Carpentras. fauna of.

Caterpillars, poisonous.

Centipedes.

Cerceris.

Chalcidia.

Chalicodoma.

Charrasse, Louis.

Chermes.

Cicada (Cigale).

Cicadelina.

Cicindela.

Cione.

Clathrix.

Clythris.

CHAPTER 16.

Clytus.

Cleona opthalmica.

Coincidence in life of parasites.

Coleoptera of Avignon.

Conchology, Fabre studies.

Copris.

Corsica.

Courrier.

Crickets, courtship of.

Crioceris.

Cuckoo.

Curves, properties of.

Darwin, Charles, Fabre an opponent of. praises Fabre. corresponds with Fabre.

Darwin, Erasmus.

Decticus.

Delagrave, Charles.

Dermestes.

Devillario, Henry.

Dorthesia.

Dufour, Léon.

Dumas.

Dung-beetles.

Duruy, Victor. sends for Fabre to attend Court. fall of.

Dyticus.

"Earth, The."

Eclipse of sun.

Education in France.

CHAPTER 16. 373

Ephippigera.

Epeïra.

Emerson.

Empusa.

Ergatus.

Eucera.

Eumenes.

Evil.

Evolution.

Fabre, Aglaë.

Fabre, Antoine.

Fabre, Antonia.

Fabre, Antonin.

Fabre, Émile.

CHAPTER 16. 374

Fabre, Frédéric.

Fabre, Henri. birthplace. childhood. boyhood. school days. a primary teacher. marriage and loss of first child. professor of physics at Ajaccio. professor at Avignon. takes up entomology. salary. poverty. as teacher. character. his pupils. goes to Court and is decorated. writes textbooks for schools. portraits of. meets J.S. Mill. denounced for subversive teaching. evicted. settles at Orange, money difficulties solved by Mill. breaks with the University. continues his series of textbooks. repays Mill money lent. dismissed from Requien Museum. researches concerning madder. leaves Orange. work at Sérignan. second marriage. his workshop. methods of work. attitude toward evolution. corresponds with Darwin. ideas as to origin of species. methods of work. compared with Réaumur. life at Sérignan. love of music. old age. poverty. jubilee celebrated.

Fabre, Henri, of Avignon.

Fabre, Jules.

Fabre, Paul.

Fabre, Mme (mother of Henri).

CHAPTER 16.

Fabre, Mme (1st wife).

Fabre, Mme (2nd wife).

Fabre, Mme Antoine.

Favier.

Female education.

Frog, bellringer.

Gadfly.

Gegner prize.

Geometry, Fabre's love of.

Geotrupes.

Glow-worm.

Goat caterpillar.

Goethe.

Grasshopper.

CHAPTER 16.

Halictus.

Harmas, the.

Heat, takes place of food.

Helix raspaillii.

Hemerobius, curious garment of.

Horace.

Horn-beetle.

Horus Apollo.

Huber.

Hugo, Victor.

Hyper-metamorphism.

Instinct.

Intelligence, function of.

Janin, Jules.

CHAPTER 16.

Jullian.

Jussieu, de.

La Fontaine.

Lamarck.

Lapalud.

Latreille.

Larra.

Leibnitz.

Leucopsis.

Libellula.

Linnaeus.

Locust.

"Log, Story of the."

Lycosa.

CHAPTER 16.

Madder, Fabre's researches concerning.

Magendie.

Malaval.

Mantis.

Maquis, the Corsican.

Marius.

Mason-bee.

Medicine, Fabre's inclination toward.

Megachile.

Meloë.

Michelet.

Mill, J.S. helps Fabre in difficulties. death of.

Mill, Mrs.

Millipedes.

CHAPTER 16.

Mimicry.

Mind, of animals.

Minotaurus.

Mistral. corresponds with Fabre.

Mitscherlich.

Montyon prize.

Moquin-Tandon.

Mushrooms, recipe for cooking.

Napoleon III.

Necrophorus.

Number, properties of. poem.

Odynerus.

Oniticella.

Onthophagus.

CHAPTER 16.

Orange, Fabre at.

Orchids, Fabre on.

"Origin of Species."

Orthoptera, primitive.

Osmia, control of sex. courtship of.

Pasteur.

Peacock moth.

Pelopaeus.

Perrier, Ed.

Philanthus.

Phryganea.

Pieris.

"Plant, The."

Pliny.

CHAPTER 16.

Poems, Fabre's.

Polygons, properties of.

Pompilus.

Potato.

Processional caterpillar.

Psyche.

Rabelais.

Raspail.

Racine.

Réaumur. compared with Fabre.

Requien of Avignon.

Requien Museum.

Rhynchites.

Ricard, Pierre, schoolmaster.

CHAPTER 16.

Rose-beetle.

Roumanille.

Saint-Léons.

Saprinidae.

Sarcophagus.

Scarabaeus sacer.

Scolia.

Scolopendra.

Scorpion.

Sérignan. Fabre settles at. evenings at.

Sicard's portraits of Fabre.

Silkworm moth.

Sisyphus.

Sitaris.

CHAPTER 16.

"Sky, The."

"Souvenirs entomologiques."

Spaeriaceae.

Sphex.

Spiders, aeronautic.

Sport, Fabre's love of.

Staphylinus.

Tachina.

Tachinarius.

Tachytes.

Tarantula.

Taylor, Harriett (Mrs. J.S. Mill).

Taylor, Miss.

Terebinth louse.

Theophrastus.

Thomisus.

Tolstoy.

Toussenel.

Trox.

Vanessa.

"Vaucluse, Flora of the."

Vaucluse, General Council of, grants Fabre a pension.

Vayssières, M.

Ventoux Alp. banquet on the.

Vezins.

Villard, Marie (Mme Henri Fabre).

Virgil.

Volucella.

CHAPTER 16.

Wasps' nest in winter.

Weevils, sloe. poplar. acorn and poplar.

Woodland bug.

Xylocopa.

End of The Project Gutenberg Etext of Fabre, Poet of Science by Legros

---

Fabre, Poet of Science by Legros

A free ebook from http://manybooks.net/